Daniela Berger

Sedimentation history along the East Greenland margin

Daniela Berger

Sedimentation history along the East Greenland margin

A geoscientific study on a continental margin in the Northern Hemisphere

Südwestdeutscher Verlag für Hochschulschriften

Impressum/Imprint (nur für Deutschland/ only for Germany)

Bibliografische Information der Deutschen Nationalbibliothek: Die Deutsche Nationalbibliothek verzeichnet diese Publikation in der Deutschen Nationalbibliografie; detaillierte bibliografische Daten sind im Internet über http://dnb.d-nb.de abrufbar.

Alle in diesem Buch genannten Marken und Produktnamen unterliegen warenzeichen-, marken- oder patentrechtlichem Schutz bzw. sind Warenzeichen oder eingetragene Warenzeichen der jeweiligen Inhaber. Die Wiedergabe von Marken, Produktnamen, Gebrauchsnamen, Handelsnamen, Warenbezeichnungen u.s.w. in diesem Werk berechtigt auch ohne besondere Kennzeichnung nicht zu der Annahme, dass solche Namen im Sinne der Warenzeichen- und Markenschutzgesetzgebung als frei zu betrachten wären und daher von jedermann benutzt werden dürften.

Verlag: Südwestdeutscher Verlag für Hochschulschriften Aktiengesellschaft & Co. KG
Dudweiler Landstr. 99, 66123 Saarbrücken, Deutschland
Telefon +49 681 37 20 271-1, Telefax +49 681 37 20 271-0
Email: info@svh-verlag.de
Zugl.: Bremen, Universität, Diss., 2009

Herstellung in Deutschland:
Schaltungsdienst Lange o.H.G., Berlin
Books on Demand GmbH, Norderstedt
Reha GmbH, Saarbrücken
Amazon Distribution GmbH, Leipzig
ISBN: 978-3-8381-1215-2

Imprint (only for USA, GB)

Bibliographic information published by the Deutsche Nationalbibliothek: The Deutsche Nationalbibliothek lists this publication in the Deutsche Nationalbibliografie; detailed bibliographic data are available in the Internet at http://dnb.d-nb.de.

Any brand names and product names mentioned in this book are subject to trademark, brand or patent protection and are trademarks or registered trademarks of their respective holders. The use of brand names, product names, common names, trade names, product descriptions etc. even without a particular marking in this works is in no way to be construed to mean that such names may be regarded as unrestricted in respect of trademark and brand protection legislation and could thus be used by anyone.

Publisher: Südwestdeutscher Verlag für Hochschulschriften Aktiengesellschaft & Co. KG
Dudweiler Landstr. 99, 66123 Saarbrücken, Germany
Phone +49 681 37 20 271-1, Fax +49 681 37 20 271-0
Email: info@svh-verlag.de

Printed in the U.S.A.
Printed in the U.K. by (see last page)
ISBN: 978-3-8381-1215-2

Copyright © 2010 by the author and Südwestdeutscher Verlag für Hochschulschriften Aktiengesellschaft & Co. KG and licensors
All rights reserved. Saarbrücken 2010

Contents

List of Figures   4

List of Tables   4

1 Introduction   5

2 Geological background   9
- 2.1 Tectonic evolution of the northern North Atlantic 9
- 2.2 Trigger mechanism for influenced cooling events 12
- 2.3 Interpretations about the glacial onset in the polar regions 13
  - 2.3.1 Antarctica (middle Eocene - early Oligocene) 14
  - 2.3.2 Norwegian margin (Pliocene) 16
  - 2.3.3 Barents Sea margin (middle Miocene - Pliocene) 16
  - 2.3.4 Southeast Greenland margin (late Miocene) 17
  - 2.3.5 Northeast Greenland margin (late Eocene - Pliocene) 17
- 2.4 Sediment transport mechanisms 18

3 Seismic data processing   23
- 3.1 Seismic reflection 23
  - 3.1.1 Multiple suppression methods 24
  - 3.1.2 Results of the multiple suppression 28

4 References   35

5 A seismic study along the Northeast Greenland margin   43
- 5.1 Abstract 43
- 5.2 Introduction 43
- 5.3 Seismic data acquisition/processing and mapping procedure 46
- 5.4 Profile description 47
  - 5.4.1 Northern Greenland Basin 47
  - 5.4.2 Southern Greenland Basin 50
- 5.5 Stratigraphy 57
  - 5.5.1 Available age information 57
  - 5.5.2 Age correlation and seismic stratigraphy 58
- 5.6 Interpretation 59
  - 5.6.1 Basement structures 59
  - 5.6.2 Total sediment thickness 62
  - 5.6.3 Glacial - Preglacial sediments 62
- 5.7 Conclusion 65
- 5.8 Acknowledgements 65
- 5.9 References 66

6 Shelf sedimentation along the East Greenland margin — 71

- 6.1 Abstract — 71
- 6.2 Introduction — 71
- 6.3 Data acquisition and processing — 73
- 6.4 ODP Site 909 — 73
- 6.5 Results — 75
  - 6.5.1 Classification of seismic units — 76
  - 6.5.2 Molloy Basin — 76
  - 6.5.3 Boreas Basin — 80
  - 6.5.4 Prograding sequences on the East Greenland shelf — 81
- 6.6 Interpretation — 83
  - 6.6.1 Sediment accumulation in the basins along the East Greenland margin — 83
  - 6.6.2 Age model of the prograding wedge along the East Greenland margin — 89
  - 6.6.3 Ice movements — 91
- 6.7 Conclusion — 93
- 6.8 Acknowledgements — 94
- 6.9 References — 94

7 Current-controlled sediments along the Northeast Greenland margin — 99

- 7.1 Abstract — 99
- 7.2 Introduction — 99
- 7.3 Oceanographic setting — 102
- 7.4 Description and interpretation of seismic reflection profiles — 103
- 7.5 Discussion — 109
  - 7.5.1 Slope sedimentation — 109
  - 7.5.2 Deep-sea sedimentation in the northern North Atlantic — 110
  - 7.5.3 Deep-sea sedimentation in the central Greenland Basin — 112
- 7.6 Conclusion — 113
- 7.7 Acknowledgements — 114
- 7.8 References — 114

8 Conclusion — 117

9 Acknowledgements — 121

List of Figures

1.1	Overview map	6
2.1	Magnetic chrons	10
2.2	Paleobathymetry 0-55 Ma	11
2.3	Climatic and Tectonic changes	12
2.4	Start of the glaciation in different polar regions	14
2.5	Glaciated margins	15
2.6	Ice sheet depositional model	19
2.7	Seismic example from the Weddell Sea shelf	19
3.1	multiple supression panel	26
3.2	f-k filtering	29
3.3	stack and predictive deconvolution	31
3.4	hyperbolic and parabolic radon transformation	32
3.5	f-k filtering	33
5.1	Location of the study area	45
5.2	Velocity model of profile AWI-20030390	47
5.3	Line drawings of the northern Greenland Basin	48
5.4	Parts of seismic depth sections in the northern Greenland Basin	49
5.5	Line drawings of the southern Greenland Basin	51
5.6	Seismic depth section of profile AWI-20030390	52
5.7	Parts of profiles in the southern Greenland Basin	53
5.8	Line drawings of the two southernmost profiles in the Greenland Basin	54
5.9	Line drawings of profiles AWI-20030585 and AWI-20030586	55
5.10	Seismic depth section of profile AWI-20030585	56
5.11	Line drawings of profiles AWI-20030145, AWI-20040230, AWI-20040231, AWI-20040232	56
5.12	Depth section of profiles AWI-20030145, AWI- 20040230, AWI-20040231, AWI-20040232	57
5.13	Stratigraphic age model	58
5.14	Magnetic anomalies in the Greenland Basin	60
5.15	Total sediment thickness grid	63
5.16	Sediment thickness grid (mid. Miocene - present)	63
5.17	Sediment thickness grid (Tertiary - mid. Miocene)	64
6.1	Map of the study area	74
6.2	ODP site 909	75
6.3	Profiles located in the Molloy Basin	77
6.4	Profiles located in the Boreas and Molloy basin	78

6.5	Profiles located in the Boreas Basin	79
6.6	Shelf profiles along the East Greenland margin	82
6.7	Map of total sediment thickness	84
6.8	Sediment thickness map of fine bedded sediments	86
6.9	Sediment thickness map of unit NA-1	88
6.10	Sediment thickness map of unit NA-2	88
6.11	Detailed bathymetry map of the East Greenland shelf	92
7.1	Location of the current-related study area	101
7.2	Deep sea channel structures	104
7.3	Downslope structures	105
7.4	Turbidity current structures	106
7.5	Well-laminated sediments	107
7.6	Fram Strait currents	108
7.7	Northern North Atlantic currents	111

List of Tables

6.1	Volume estimations	85

1 Introduction

Presently, the Northern Hemisphere is dominated by a cold climate and sea ice depending on the season. Greenland, the world largest island, has a relatively constant ice coverage. The total land area amounts to 2166086 km^2, of which the Greenland ice sheet covers 1755637 km^2 (81 %). The weight of the ice has depressed the central land area into a basin shape lying more than 300 m below the sea level, so that in general the ice drains to the coast from the center of the island. The topography, especially on the east coast is regionally very rough and steep.

Geoscientific operations in the shallow marine part off East Greenland are logistically difficult to conduct, because of the heavy ice conditions. For the same reason, our geological knowledge about the Northeast Greenland shelf is very sparse. Access to the Norwegian margin is far simpler, because the climate is influenced by the warm Gulf stream, which prevents ice from building up.

It is of interest to understand the evolution of both margins. Geophysical investigations in East Greenland started 120 years after the famous Koldewey expedition in 1869-1870 with the research vessel "Germania". The first geophysical measurements were carried out in the early 1980's on the Southeast Greenland shelf (Larsen et al., 1990) in the form of seismic reflection data. Seismic refraction investigations were made for the first time 1988 by the University of Hamburg and Alfred Wegener Institute for Polar and Marine research (Weigel et al., 1995). Most of the previous studies on the East Greenland continental margin had been concentrated on the volcanic history and the deep crustal structure (Schlindwein and Jokat, 1999; Voss and Jokat, 2007).

Studies at the conjugate margin of Norway are relevant for understanding the opening of the Arctic gateway, differences in the glaciation history and sediment transport processes. Especially the Fram Strait (Fig. 1.1), the deep-water connection between the North Atlantic and Arctic Ocean is of interest to oceanic circulation issues. Until recently it has been unclear, how important this gateway has affected the global climate change (Jacobsson et al., 2007). Clearly the opening of the Fram Strait is important since the thermohaline circulations are the strongest regulator of the global climate.

The lack of bore hole information on the NE Greenland shelf in contrast to the well, sampled Norwegian margin, in combination with few seismic reflection data on the shelf, make it difficult to provide constraints about the glaciation history on the East Greenland shelf. Until recently it was not possible to correlate the stratigraphic information of ODP site 913 (Greenland Basin), ODP site 908 (on the Hovgård Ridge) and ODP site 909 (north of the Hovgård Ridge) with that of the adjacent NE Greenland shelf, because of missing seismic data.

However, seismic data have been acquired by AWI during summer campaigns (1997, 1999, 2002, 2003) and these data have enabled tying the mentioned ODP sites to the NE Greenland shelf. In total, over 10000 km of seismic reflection data were gathered with a 600 m and 3000 m long streamer. Additionally, ocean bottom seismometer and sonobuoy data were acquired to map the deep structure of the East Greenland continental margin as well as the transition to the oceanic crust. The profiles are located between 72°N and 82°N (Fig. 1.1).

The main questions the surveys should answer are:

1. How did the passive continental margin of East Greenland develop after the opening of the North Atlantic?

Introduction

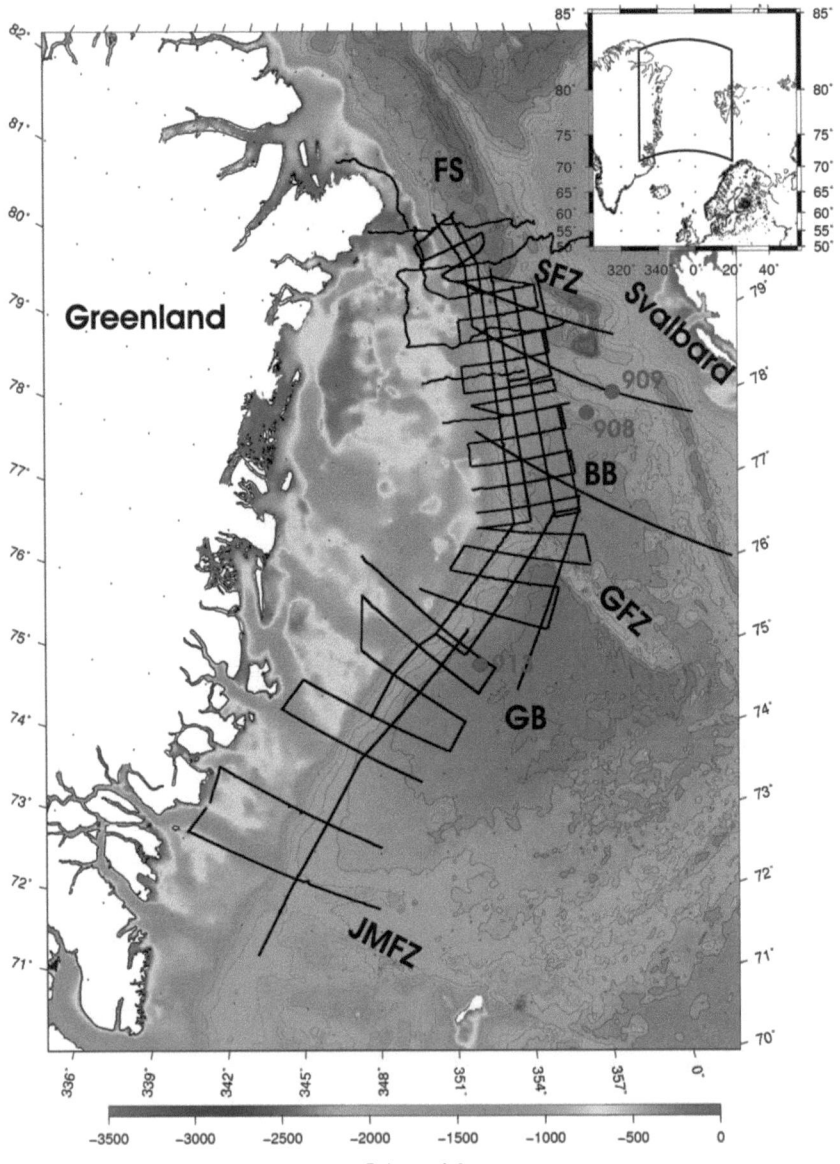

Figure 1.1: Overview map of the study area with the international bathymetric chart of the Arctic Ocean (IBCAO) in the background (Jakobsson et al., 2001). The black lines represent the AWI seismic reflection profiles along the Northeast Greenland margin. The red dots demonstrate the locations of the ODP drill sites. FS = Fram Strait, HR = Hovgård Ridge, SFZ = Spitsbergen Fracture zone, MB = Molloy Basin, BB = Boreas Basin, GFZ = Greenland Fracture zone, GB = Greenland Basin, JMFZ = Jan Mayen Fracture zone

2. How often did the ice sheets advance to the shelf break and is it possible to give an estimate about the onset of the glaciation through the age correlation?

3. Are there lateral variations in the sedimentation along the East Greenland margin, and if so, what are the influencing factors?

4. Can seismically observable sediment structures give information about current activity and climate changes in the region investigated?

The new data base allows for the first time an extensive analysis of the sedimentary structures along the East Greenland margin. In conjunction with several Ocean Drilling Program (ODP) legs this thesis enhances our understanding of glaciation on East Greenland.
The thesis is structured as follows:

Chapter 1 gives a short introduction into the problem of the geological interpretation of the East Greenland margin to emphasize why scientific investigations are important in this area. The main objectives of the surveys are listed.

Chapter 2 is focused on the geological evolution of the Arctic region, the glacial history of the polar regions, and the glacially related sediment transport.

Chapter 3 explains the applied processing steps with the focus on multiple suppression methods.

Chapter 4 presents: *Berger and Jokat, 2008. A seismic study along the East Greenland margin between $72°N$ and $77°N$. Geophysical Journal International.*
This study is focused on the interpretation of the basement structure and an age correlation between sedimentary units in the Greenland Basin and the adjacent shelf region.

Chapter 5 presents: *Berger and Jokat, re-submitted March 2009. Sediment deposition in the northern basins of the North Atlantic and characteristic variations in shelf sedimentation along the East Greenland margin. Marine and Petroleum Geology.*
This study is focused on sediment depositions in the Boreas, Molloy and Greenland basins. Seismostratigraphy along the Northeast Greenland shelf and variations in the shelf sedimentation have given reason to assume different sediment transport processes.

Chapter 6 presents: *Berger and Jokat, submitted March 2009. Current-controlled sedimentation along the Northeast Greenland margin. Marine Geology.*
This study is focused on current-controlled sediment transport processes and their accumulation structures in conjunction with climate changes.

Chapter 7 summarizes all the results of the thesis and gives an outlook for further studies.

2 Geological background

2.1 Tectonic evolution of the northern North Atlantic

Excluding Late Devonian post-orogenic extensional collapse of the Caledonian Orogeny, rifting that eventually led to break-up of the NE Atlantic and Arctic started in Carboniferous time (Moore et al., 1992). The late Permian to Middle Jurassic initial phase of Pangaea breakup invoked the southward and westward propagation of the proto-Norwegian-Greenland Sea (NGS) and the Tethys rift systems, whereas the late Jurassic to Paleogene break-up phase was dominated by the stepwise northward propagation of the central Atlantic (Ziegler, 1990). The Central Atlantic opened in Middle Jurassic time (175 Ma) and was followed by northward propagation of the spreading system between Iberia and Newfoundland in Aptian/Albian time (e.g. Tucholke, 2007), while rifting intensified in the proto-Norwegian-Greenland Sea (Ziegler, 1990). Seafloor spreading opened the Labrador Sea in Paleocene time (Chalmers and Pulvertaft, 2001), and shifted to the east side of Greenland in Early Eocene time when the NE Atlantic opened (e.g. Talwani and Eldholm, 1977). The Eurasia Basin represents the northern tip of the Atlantic rift system and this ocean started spreading in Late Paleocene time or Early Eocene time (Brozena et al., 2003). The Fram Strait represents the only deep water connection between the Arctic Ocean and Atlantic Ocean. A reconstruction of this gateway and the evolution of the basins in the North Atlantic is important for the interpretation of a correlation to climate changes. A recent interpretation of the opening of the northern North Atlantic is shown by Engen et al. (2008), Knies and Gaina et al. (2008) and Ehlers and Jokat (2008). A detailed study of the tectonic opening history is based on a compilation of different geophysical data: magnetic data (Fig. 2.1), seismic data, Arctic drill hole information, bathymetric data, and crustal thickness (Ehlers and Jokat, submitted 2008a). The continental drift from 55 Ma up to the present scenario is imaged in 5 Myr steps (Fig. 2.2). The data analysed within Ehlers and Jokat's study are located between 65°N and 81°N. The oldest sea-floor spreading anomaly (C24b; 56Ma) was identified in the Norwegian Greenland Sea (Talwani and Eldholm, 1977), the area where the separation of East Greenland and Norway started. The southern boundary of the Norwegian-Greenland Sea is defined by a shallow transverse aseismic ridge between Greenland, Iceland, and Scotland. The region can be divided into three areas separated by fracture zones (Fig. 1.1). The southern region lies between the Greenland-Scotland-Ridge and the Jan Mayen Fracture Zone. The central part (Greenland Basin) is bounded by the Jan Mayen and Greenland Fracture zones, whereas the northern region (Boreas and Molloy basins) lies between the Greenland-Senja and Spitsbergen fracture zones (Talwani and Eldholm, 1977; Myhre et al., 1995). The area between the Greenland Sea and Arctic Ocean is not known in detail, largely due to the lack of modern aeromagnetic data (Fig. 1.1). In the model (Fig. 2.2) of Ehlers and Jokat (2008), the oldest basin between 65°N and 81°N is the Greenland Basin which these authors give an Early Eocene age of 56 Myr. Furthermore, we can observe the beginning of the seafloor spreading in the Boreas Basin at 38 Ma (Fig. 2.2), and due to the late breakup of the Molloy Basin (21 Ma) this is the shallowest basin in the north (Ehlers and Jokat, 2008). The opening of the Fram Strait is now estimated to the Middle Miocene based on results from ACEX borehole 302 (Jakobsson et al, 2007) but also on plate reconstructions (e.g. Engen et al, 2008). Ehlers et al. (submitted 2008b) modeled deep-water exchange (Figs. 2.1, 2.2) between the Arctic Ocean and the North Atlantic Oceans to the Middle Miocene (17-18 Ma), similar to the proposed age by Jakobsson et al. (2007).

Geological background

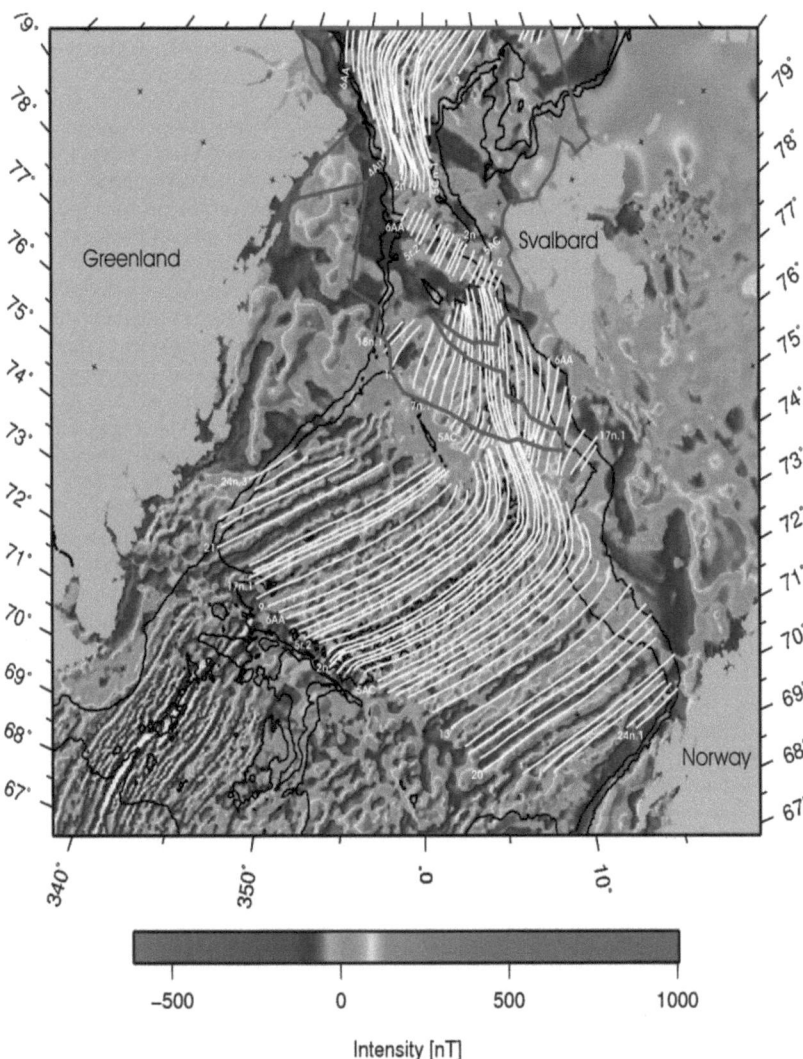

Figure 2.1: Identified seafloor spreading anomalies by Ehlers and Jokat (submitted 2008a). The magnetic details derived from the GAMMA5 grid (Verhoef et al., 1996) and extended by new aeromagnetic AWI data between 75°N and 85°N framed in red (Leinweber, 2006). The magnetic anomalies were labeled after Gradstein et al. (2004).

2.1 Tectonic evolution of the northern North Atlantic

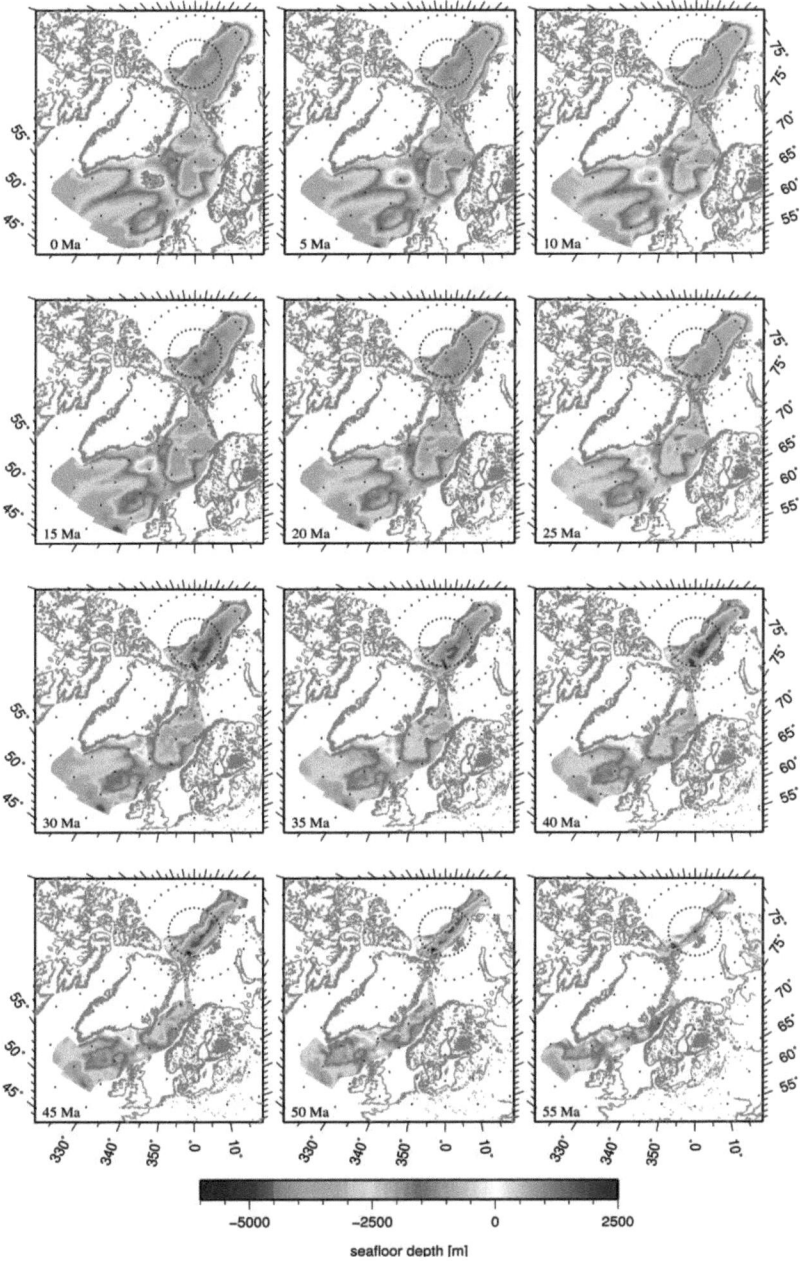

Figure 2.2: Modelled paleobathymetry for the northern North Atlantic from Ehlers and Jokat (submitted 2008a) for 0-55 Ma.

2.2 Trigger mechanism for influenced cooling events

Investigations of deep sediment cores provide important input to understanding palaeoclimate changes. Zachos et al. (2001) have compared oxygen ($\delta^{18}O$) isotopes, climatic and tectonic events for the period ranging from 65 Ma to present (Fig. 2.3).
The study showed that during the last 65 million years, Earth's climate system has experi-

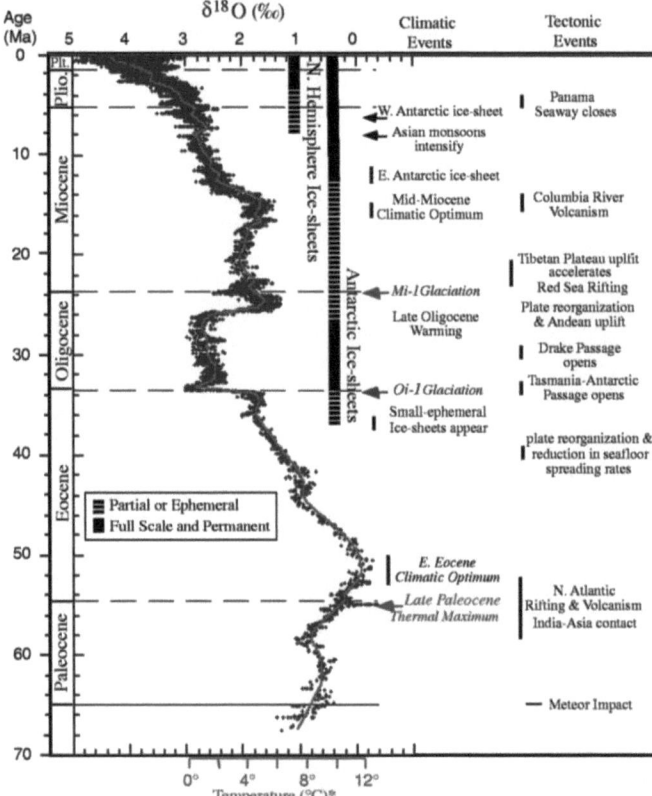

Figure 2.3: Table of climatic and tectonic events, $\delta^{18}O$ and temperature after Zachos et al. (2001).

enced continuous change. Climate changed from extremes of expansive warmth with ice-free poles, to extremes of cold with massive continental ice-sheets and polar ice caps (Zachos et al., 2001). Some tectonically driven events can be interpreted to have triggered major shifts in the dynamics of the global climate system.
Opening of the NE Atlantic and collision of India-Asia are commonly attributed with a decrease of $\delta^{18}O$ isotopes, reflecting an increase of the deep-sea temperature. The end of the super-greenhouse period that had started in Creaceous time is known as early Eocene Climatic Optimum (EECO; 52 to 50 Ma) with deep-sea temperatures of +12°C. After this maximum a 17 Myrs long cooling trend led to the first partial and ephemeral ice formations

in Antarctica. In early Oligocene times (∼34 Myr) the separation of the African and Australian continents from Antarctica (Tasmania-Antarctic Passage) contributed to the onset of the Antarctic Circumpolar Current (ACC). Approximately 3 Myrs later, the opening of the Drake Passage enabled an eastward flow of the ACC (Lawver and Gahagan, 2003). A following cooling trend (Fig. 2.3) resulted in a rapid expansion of the Antarctic continental ice sheet during the early Oligocene. These temperature conditions led to permanent ice coverage in Antarctica (Zachos et al. 2001) up to the late Oligocene onset of warming (26 to 27 Ma). This following warm period until middle Miocene times reduced the extend of Antarctic ice. The global bottom water temperature proxies show a slight increase (Wright et al., 1992). This phase of warming peaked in the Mid-Miocene Climate Optimum, and was followed by a gradual cooling and re-establishment of a major ice-sheet on Antarctica (Fig. 2.3). Lawver and Gahagan (1998, 2003) suggested that the collision of the Australia-New Guinea block with Southeast Asia around 15 Ma was an important tectonic event with respect to ocean circulation. Among other global causes, they indicated that the blocking of substantial equatorial transport of water from the Pacific into the Indian Ocean enhanced the driving forces of the Antarctic Circumpolar Current, and led to the middle Late Miocene expansion of the East Antarctic ice-sheet. In the Northern Hemisphere the deep water connection between the Arctic ocean and North Atlantic (Fram Strait) opened also in Middle Miocene times (Jakobsson et al., 2007; Ehlers and Jokat, 2008). The first deep-water exchange took place (Jakobsson et al., 2007). The influence of this gateway opening on climatic changes and the onset of glaciation is sparsely investigated. The continuous increasing of $\delta^{18}O$ isotopes and decreasing of bottom water temperature in the world's oceans resulted in a permanent ice coverage in Antarctica, and led to first partial and ephemeral ice formations in the north around 12-14 Ma (Darby and Zimmermann, 2008) (Fig. 2.3). The early Pliocene is marked by subtle warming trend until ∼3.2 Ma, when $\delta^{18}O$ again increased reflecting the onset of Northern Hemisphere glaciation (Shackleton et al., 1988). In figure 2.3 (Zachos et al., 2001) the closure of the Panama seaway should have an influence on the climate and ice formation in the north. The Gulf Current have been shifted towards north, resulting in more rainfall/snowfall and ice concentration in the Arctic.

2.3 Interpretations about the glacial onset in the polar regions

A main part of this study was to investigate the glacial structures on the Northeast Greenland shelf. On the basis of ODP site 913 (Greenland Sea) and site 909 (north of the Hovgård Ridge) a first age model based on seismostratigraphy could be determined for sediments on the Northeast Greenland shelf (see chapter 4 and 5). The glacial sediments look similar to glacial sediments in other polar regions. Hence, this section gives a short overview about the different interpretations of the glacial onset in selected polar regions. In figure 2.4, the onset of glaciation in Antarctica, Norwegian margin, Barents Sea margin, Southeast Greenland margin and Northeast Greenland margin is summarized. The various assumption for the start of the glaciation in a region is marked by the hatchured rectangles. The upper boundary of the rectangles shows the youngest age of the glaciation and the lower boundary represents the interpreted onset of glaciation for the selected region. A more detailed description giving the basis of the various ages for the onset of glaciation is provided in the following section and is ordered by region. The rectangles on figure 2.5 demonstrate the location of former and presently glaciated margins discussed in the text. The red highlighted box represents the region of the Northeast Greenland shelf, where this thesis concentrates. Additionally, the locations of the ODP drill sites cited in the text are marked as black dots labelled according

Geological background

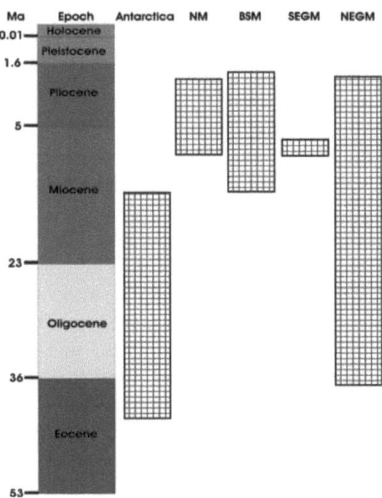

Figure 2.4: Start of glaciation in different polar regions. The rectangle infill demonstrates the range of the different interpretations for the onset of glaciation in this region. NM = Norwegian margin, BSM = Barents Sea margin, SEGM = Southeast Greenland margin, NEGM = Northeast Greenland margin.

to the drill site number in this figure.

2.3.1 Antarctica (middle Eocene - early Oligocene)

About 90 % of the global ice mass is accumulated on Antarctica (Denton et al., 1991) and around 98 % of the Antarctic continent is covered by an ice sheet averaging 1 to 4 km in thickness (Nielsen et al., 2005). Parts of the continental shelves are also covered by large ice sheets (Anderson, 1999).
Since several years, bathymetric data, multichannel seismic data and sediment core data have been collected and these data provide a detailed view of the shallow parts of the sedimentary section. The data show evidences for ice dynamic and indicate that grounding and eroding ice streams developed on the continental shelf. The numerous data were acquired in different regions around Antarctica (Fig. 2.5). The following subitems demonstrate the various estimations for the onset of glaciation in that region.

- middle Eocene or earliest Oligocene for Prydz Bay (Hambrey et al., 1989): based on drilling results of ODP site 740 and site 741 tied into a seismic network

- 30 - 33.42 Ma for the Wilkes Land margin(De Santis et al., 2003; Escutia et al., 2005): based on extrapolation of sedimentation rates from ODP site 269 (400 km north of the shelf) to the middle to lower continental rise

- early Miocene for the Ross Sea (De Santis et al., 1995): based on a large MCS data set combined with age information from ODP site 272 and 273

- early Oligocene for the western Ross Sea (Brancolini et al., 1995): based on borehole information from ODP site 270

2.3 Interpretations about the glacial onset in the polar regions

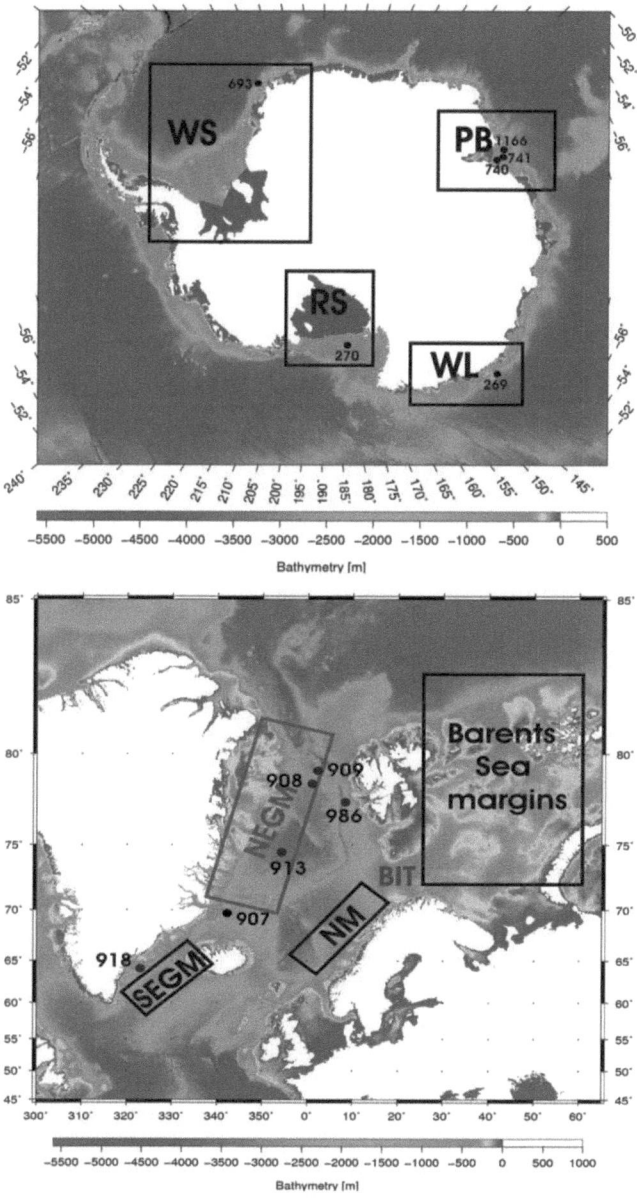

Figure 2.5: The upper image demonstrates four glaciated margins in Antarctica: PB = Prydz Bay, WL = Wilkes Land, RS = Ross Sea, WS = Weddell Sea. Below adjacent glaciated margins to the Northeast Greenland margin are shown (NM = Norwegian margin, BSM = Barents Sea margin, SEGM = Southeast Greenland margin, NEGM = Northeast Greenland margin). The black dots represent the locations of drill sites.

Geological background

- Early to late Oligocene for Weddell Sea/Dronning Maud Land (Miller et al., 1990): based on MCS data and ODP drill hole information from site 692 and 693

2.3.2 Norwegian margin (Pliocene)

The Norwegian margin (Fig. 2.5, NM) is one of the best explored margins in the Northern Hemisphere. Even so it can be claimed that, the Cenozoic glacial history is not fully understood. In general, the onset of large-scale progradation is interpreted to mark the beginning of grounded ice sheets advanced towards the shelf break. The different estimations for this event are:

- 2.56 Ma (Jansen et al., 1988): based on samples from exploration wells off Mid-Norway and ice rafted debris

- 2.74 Ma (Jansen et al., 2000): the authors revised their previous assumptions (1988) because of a revised time scale

- older than 2.3 Ma (Eidvin et al. 1998, Rise et al., 2005): based on planktonic fossil fauna correlated with palaeomagnetically calibrated fossil zones from ODP/DSDP drill sites in the Norwegian Sea on exploration wells

- 2.53 Ma (Wagner and Hölemann, 1995): based on analyses of organic geochemical data from different ODP sites on the Vøring Plateau

- 3.5 Ma (Jansen et al., 2000): based on a continuous IRD record produced from ODP site 907

2.3.3 Barents Sea margin (middle Miocene - Pliocene)

The Barents Sea Shelf (Fig. 2.5) represents the wide shelf region extending up to 82°N. In the southwestern part of the Barents Sea the large Bear Island Trough (BIT) is distinctly visible in the bathymetry (Fig. 2.5). This bathymetric trough, leads to a large trough mouth fan off the SW Barents Sea margin. This sediment deposition area is well investigated by seismic, measurements but also borehole information are available. ODP site 986 on the west Spitsbergen margin is a key borehole for the evolution of the Barents Sea shelf and the sediment transport in this region (Forsberg et al., 1999; Channell et al., 1999). The different age models for the onset of glacial erosion are:

- 2.3 Ma (Forsberg et al., 1999) - 2.58 Ma (Channel et al., 1999): correlation of a regional unconformity (R7) traced along the Barents Sea- and Spitsbergen margin to ODP site 986

- 2.20 - 2.35 Ma (Sættem et al., 1992): based on $^{40}Ar^{39}Ar$ dates on cored volcaniclastic material immediately underlying the wedge sediments in the Vestbakken volcanic province of the Bear Island Fan

- middle Miocene (Knies and Gaina, 2008): data from ODP site 909 on the Hovgård Ridge, as well as preglacial paleorelief and bathymetric reconstructions in the Barents Sea point to large-scale glaciations developed in the northern Barents Sea

2.3.4 Southeast Greenland margin (late Miocene)

The first seismic reflection measurements offshore Greenland were aquired in the early eighties on the Southeast Greenland margin (Larsen et al., 1990). Former studies were concentrated onshore, and little was known about the onset of the glaciation on the conjugated margin of Norway. These profiles were used as pre-site surveys for drilling campaigns. During September-November 1993, core driling at different drill sites were carried out (Leg 152, Larsen et al., 1994) on the Southeast Greenland margin. One of the scientific objectives of Leg 152 was to investigate glaciomarine processes and history of the Southeast Greenland margin (Larsen et al., 1994). The different age models for the onset of glacial erosion based on this scientific cruise are:

- 7 Ma (Larsen et al., 1994): based on nannofossil biostratigraphy and sedimentation rates on ODP site 918

- 11 Ma (Helland et al., 1997): analyses of samples from ODP site 918 (Sr-isotopes from planktonic foraminifers)

2.3.5 Northeast Greenland margin (late Eocene - Pliocene)

Because of the heavy drift ice conditions along the Northeast Greenland shelf it is difficult to carry out seismic measurements there and even more problematic to pursue drilling projects. The northernmost ODP drill site 913 (Leg 151) along the Northeast Greenland margin (Fig. 2.5, NEGM) is situated in the deep Greenland Basin just off the Northeast Greenland shelf (Myhre et al., 1995). Until recently, missing profiles from the deep sea onto the shelf complicated the interpretation of the onset of glaciation on the Northeast Greenland margin. Existing interpretations are based on reflection character, regional considerations, onshore Greenland geology and results from borehole samples of drill site 913. The first seismic measurements on the Northeast Greenland shelf were carried out north of 79°N by Hinz et al. (1991) in the year 1988. The results were interpreted to reflect a very deep Permo-Carboniferous rift basin filled with sedimentary rocks beneath the inner shelf. Between 1992 and 1995 the seismic network was extended to the south by a group of industrial industries. A general overview of this seismic campaign and of the geological history of the Northeast Greenland shelf was published by Hamann et al. (2005). Other studies on this margin (Hinz et al., 1987) were concentrated on the deep sea and slope area on the volcanic history along the Northeast Greenland margin. The various assumptions for the onset of glaciation are:

- Plio - Pleistocene (Hamann et al., 2005; Tsikalas et al., 2005): subdivision of the offshore sediment succession into several units dated by comparison to the onshore successions of central East Greenland and Northeast Greenland, as well as offshore successions in the southern Barents Sea and on the Mid-Norwegian margins

- 30 Ma - 38 Ma (Eldrett et al., 2007): based on biostratigraphy and samples from ODP site 913, results show evidences of ice-rafted debris including macroscopic dropstones

- middle Miocene (Helland & Holms, 1997; Winkler et al., 2002): based on proxy data and ice-rafted debris (IRD) from the Nordic seas

None of the above results were based on a seismostratigraphic tie to ODP site 913 in the deep Greenland Basin. The AWI (Alfred Wegener Institute) data, collected during the summer seasons 1997, 1999, 2002 and 2003 made it possible to tie the shelf succesion with ODP site 913 and hence to investigate the sediment structure along the Northeast Greenland shelf, slope and deep sea region. The results are the basis of this thesis and will be interpreted in the chapter 5, 6 and 7.

2.4 Sediment transport mechanisms

Sparse 2D seismic data cannot be used to prove occurrences of glacial detritus, but provide a compelling evidence of an abundant sediment supply from the continental shelf to the deep-marine environment. This sedimentary supply can be attributed likely to ice sheet transport mechanisms, as discussed below. Different sedimentary processes have been observed on glacially influenced margins, mostly documented along the well investigated Antarctic margin. This chapter gives an introduction into sediment transport mechanisms observed on the East Greenland magin (see chaper 4, 5 and 6). The main transport mechanisms are:

- Sediment transport by ice sheets
- Sediment transport by ice streams
- Sediment transport by icebergs
- Sediment transport by gravity-driven process
- Sediment transport by current activity

Sediment transport by ice sheets Geological and geophysical data were acquired on various continental shelf areas (e.g. Antarctica, Barents Sea, Southeast Greenland) showing evidence for repeated ice advances and changes in sediment accumulation. The advances of grounded ice in times of glacial maximum transported large volume of sediment across the shelf edge and are regarded as the mechanism for the development of prograding clinoforms. A model of ice sheet deposits is image in figure 2.6 and figure 2.7 shows an example of a seismic profile from the Weddell Sea shelf (Fig. 2.7) with the characteristic geometry of the prograding sequences. These structures could be also recognized on the Northeast Greenland shelf and points to an glacially influenced shelf region. The results along the East Greenland shelf are shown and interpreted in chapter 4 and 5.

Changes in the geometry of the prograding sequences from low-angle strata to steep foreset strata are likely the result from changes in the glacial processes (Escutia et al., 2005). A major change in sequence geometry (start of progradation) on the outer shelf of the western Antarctic Peninsula was dated and interpreted as the start of advances of grounded ice to the shelf break (Cooper et al., 1991). In interglacial times, biogenic material accumulated above the older sediment deposits. The repeated advance of glaciers over areas of a nearly flat relief, carries debris from its usually more elevated source area and takes up further material by basal erosion. If the base of the glacier is cold and frozen to the ground, more basal debris is generally accumulated than by "temperate" glaciers, which have a wet base sliding relatively easily over the substrate. Melting of glacier ice at its base leads to the release and accumulation of debris in the form of melt-out till (Einsele, 1992).

2.4 Sediment transport mechanisms

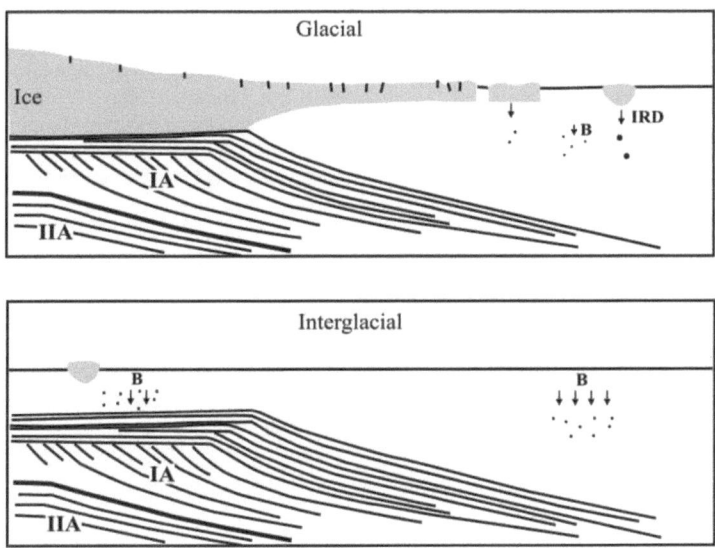

Figure 2.6: Ice sheet depositional model of Cooper et al., (1991), sequences of Unit IA are interpreted as prograding clinoforms and sequences of Unit IIA are interpreted as aggrading clinoforms. B=Biogenic, IRD=Ice rafted debris.

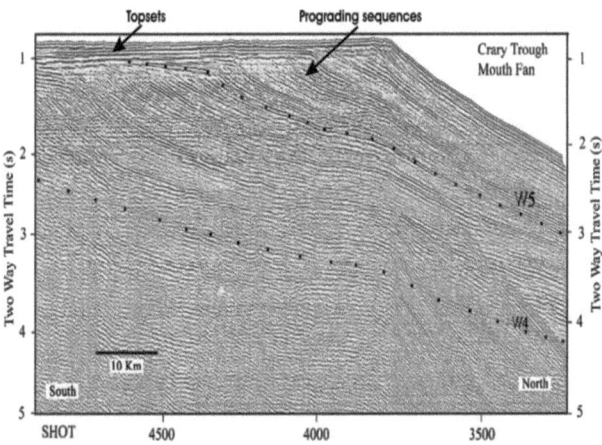

Figure 2.7: Seismic reflection profile from the Weddell Sea shelf in the prolongation of the Crary Trough taken from Kuvaas and Kristoffersen (1991). Topsets and prograding strata are shown in the outer shelf area like on the Northeast Greenland shelf (therefor see chapter 4 and 5.

Geological background

Sediment transport by ice streams A strong cyclicity in sediment supply on the outer shelf and upper slope region can be related to periods when the grounding line reaches the shelf edge but very little otherwise. Therefore, the advanced glaciers show a bulldozing effect at the front of grounding ice. Unsorted sediment (till) is transported beneath ice streams and deposited directly in front of the grounding line (Barker, 1995). When grounded ice streams extend onto the continental shelf, massive to crudely stratified diamictons are deposited in the shelf troughs (e.g. Wilkes Land). Diamictons were deposited as tills (imaged in seismic lines as topsets; see also figure 2.7) directly below the grounded ice sheet (Anderson et al., 1980; Domack, 1982) or as glacial open-marine diamictons opposite to the ice stream terminus as the foresets of the prograding wedges, also known as trough mouth fans (Escutia et al., 2005). Such sediment deposits can be found at the Scoresby Sund (Dowdeswell et al., 1997), Prydz Bay Channel (Kuvaas and Kristoffersen, 1991) and Bear Island Fan (Laberg and Vorren, 1996). The records with highest resolution of Cenozoic glaciation on the Antarctic continental margins are found in thick sediments of trough mouth fans, situated in front of deep bathymetrical troughs. Large glacial troughs are also visible along the entire East Greenland margin. Compiled sediment thickness maps (see chapter 5) give information about the amount of sediment material accumulated in the prologation of glacial troughs in comparison to the adjacent areas.

Sediment transport by icebergs In regions of very cold climate and sufficient ice accumulation, large areas can be covered with comparatively thick shelf ice. Most of the debris in these floating ice sheets is located near their base (Einsele, 1992). In this environment, proglacial marine sediments are often rich in diatoms, bioturbated, and contain few dropstones. Areas with large and "dirty" valley glaciers reaching the sea can provide great amounts of siliciclastic debris. Icebergs originating from such dirty glaciers frequently carry their debris load over long distances, and hence are the source of dropstones found in pelagic and hemipelagic sediments in large ocean basins far away from glaciated regions. Such sediment components were found in ODP site 909 (north of the Hovgård Ridge) and site 913 (deep Greenland Basin). Myhre et al. (1995) and Knies and Gaina (2008) believe in sediment transport by icebergs from the Barents Sea shelf towards the Fram Strait up to the area where site 909 is located. Dropstones in core samples from site 913 were assigned to sediments eroded from the Greenland shelf. The seismostratigraphic interpretation on the basis of these two bore holes will be introduced in chapter 4 and 5.

Sediment transport by gravity-driven processes For sediments in the upper slope region mass transport by gravity-driven processes plays an important role. These gravity flows can be generated with minimal triggering on steep slopes. Sediment overloading at the shelf break during times of glacial maxima and earthquakes can trigger instability and generate slumps. It has also been postulated that, at glaciated margins, such as the Labrador Sea, gravity flows can be generated by direct bedload-rich melt-water discharges from the glacier terminus sides (Hesse et al., 1997). Large volumes of sediment were transferred from the slope into deep basins. Numerous large turbidity current channels, which cut into the lower slope and rise, are identified along continental margins (Dowdeswell et al., 2004). Results from the analyzed data along the Northeast Greenland margin will be discussed in chapter 6 with regard to sediments transported by gravity-driven processes. Surface sediments from the slope environment consist of massive sandy mud to fine sand, alternating with massive sandy and pebbly deposits (Domack, 1982; Escutia et al., 2003), which resemble the diamictons of the continental shelf (Escutia et al., 2003). Base of slope and upper continental rise sedimentation is characterized by overall finer-grained deposits (e.g., mud and silty to sandy mud). At the Antarctic continental margin the isostatic rebound is suggested to have generated gravity flows (Anderson et al., 1979; Escutia et al., 2005).

Sediment transport by current activity Another sediment transport medium are currents with a strong erosional character. They lead to widespread deposition of contourite drifts and formation of channel-levee systems. Such accumulations can be observed on different margins e.g. Antarctica (Kuvaas et al., 2004), South Africa (Schlueter and Uenzelmann, 2007), the northwestern Sea of Okhotsk (Wong et al., 2003), and the NE Atlantic (Stoker et al., 2005; Laberg et al., 2005; Hjelstuen et al., 2007). The determination of the age of drift bodies can point to the onset of major new ocean current circulation systems. The constitution of channel-levee systems give details about the flow directions. Hence, it has been possible to propose a link between the opening of gateways, like the Drake Passage in the Southern Hemisphere and contourite deposits. The Circum-Antarctic current is a popular explanation for the earliest Oligocene global cooling. After the opening of the Fram Strait and the deep-water exchange between the North Atlantic and Arctic Ocean contourites and drift deposits could be developed (see chapter 6).

3 Seismic data processing

3.1 Seismic reflection

This study includes seismic data acquired during different AWI expeditions along the East Greenland margin. The dataset recorded during the summer season 2003 was processed and interpreted. For a detailed interpretation along this margin, I also used the processed datasets acquired in 1997 and 2002. The dataset carried out in the Greenland Basin (Fig. 1.1) was recorded with different streamer lengths, depending on sea ice coverage during this campaign. A 600 m long streamer was used in the north of the Greenland Basin (75.5°N-77°N) whereas south of 75.5°N a 3000 m long streamer was used. The seismic energy was consistently generated with an array of eight airguns with a total chamber capacity of 24 litres. The shot interval was 15 s and the sampling interval 2 ms.

The following workflow shows the applied processing steps:

Demultiplexing

⇓

CDP sorting

⇓

Editing

⇓

Filtering

⇓

Spherical divergence correction

⇓

Velocity analysis

⇓

Normal moveout correction

⇓

Multiple suppression

⇓

Stacking

⇓

Depth conversion and interpretation

Time sequences from the field data tapes are resorted to short gathers for each shot. The data were tied to the navigation data (latitude, longitude, course, cruising speed and water depth). With an average cruising speed of 5 knots and a shot interval of 15 seconds, the shot point spacing is approximately 37 m. The shot gathers were sorted into CDP (common depth point) groups. For both streamer configurations (streamer lengths: 600 m and 3000 m) we chose a CDP spacing of 25 m.

After CDP sorting the data were edited, whereby noisy signals and bad/dead channels were removed and wrongly polarized traces were corrected. A band pass filter of 12 - 80 Hz was applied and frequencies below 12 Hz and above 80 Hz were muted out. To create a representative velocity model, two different velocity analysis methods are available: interactive velocity analysis and constant velocity analysis. Especially in the shelf region with strong seafloor multiples, a constant velocity analysis was applied to obtain the best quality stack of signal. With these analyzed velocities the normal moveout correction was applied and the data were stacked.

Special multiple suppression methods will be described. To convert the time sections into depth sections, interval velocities were taken from the 3000 m long streamer data. Additionally, velocities from seismic refraction data from Voss and Jokat (2007) have been used. For profiles north of line 20030390, I have interpolated the interval velocities of this line to determine uniform seismic depth sections for the entire network.

Finally, the navigation of the profiles and the seismic data were loaded into the seismic interpretation system LANDMARK.

3.1.1 Multiple suppression methods

Multiples are secondary reflections with interbed and intrabed raypaths. Guided waves include supercritical multiple energy (Yilmaz, 2001). For special travel times they can interfer with the primary signal resulting in a poor signal to noise ratio. We can differenciate between multiples with short and long wave paths. Multiples with short wave paths (peg-leg) partly interfere with primary signals and are difficult to observe. Much more annoying are multiples with long wave paths (e.g. seafloor multiples), because they often mask the complete subsurface structure.

It is essential to suppress seafloor multiples. Actually, after stacking, seafloor multiples should be removed. Incoherent signals (multiples) will be suppressed and coherent signals (primary signals) will be intensified. That presumes, the velocity gradient is great enough to distinguish between seafloor multiple and real seafloor reflection. Especially, in shallow water depths (shelf region) on the East Greenland shelf the multiples are very strong, because of the high acoustic impedance contrast (density multiplied with velocity) which leads to high velocities near the subsurface. The high velocities (around 2300 m/s in the shelf region) can be related to the strongly influenced glacial shelf (discussed in chapter 5). Because of the thick ice coverage on the shelf the glacial material has become over-compacted.

To get a representative model for the subsurface the following multiple suppression methods were tested:

a) Predictive deconvolution
b) Radon transformation in the τ-ρ-domain (hyperbolic/parabolic)
c) f-k filtering.

The results will be discussed after a short explanation of the multiple suppression methods. The following description should pinpoint to the different mathmatic approaches. Complete

//
and detailed explanations are given e.g. by Militzer & Weber (1987), Sheriff & Geldard (1995) and Yilmaz (2001). Within the appendix, the processing steps with all used parameters are listed.

a) Predictive deconvolution

In general, the aim of the deconvolution is an improvement of the resolution. The ideal case would be a spike signal, which produces a broad frequency spectrum. The deconvolution is an inverse filter and with a predictive deconvolution it is possible to suppress alternate, circular signals, like seafloor multiples. Prior to the application it is necessary to determine a function of autocorrelation to define the operator length (length of the signal to create a spike signal). Additionally, the prediction length specify the time from multiple to multiple signal. For a successfully execution the following requirements have to be compiled (Yilmaz, 2001):

- static source signal
- minimum-phase signal
- subsurface consist of horizontal layers with constant velocity
- no noise signals within the data

The reality shows, that these requirements are not fulfilled. Therefore, numerical noise in the seismogram is generated.
The first panel in figure 3.1 shows the result of the predictive deconvolution. The first multiple is visible at around 780 ms. The amplitude spectrum below exhibits an attenuation of the multiple of 5 db (from -4 db of the seafloor reflection to -9 db of the first multiple) in comparison to the primary signal.

b) Radon transformation in the τ-ρ-domain

In the τ-ρ-domain we can distinguish between two kinds of transformation: hyperbolic and parabolic.

hyperbolic This technique uses a hyperbolic curve to sum the samples of incoming (x,t) and outcoming (v,τ) coordinates. The hyperbola is defined as:

$$t^2 = \tau^2 + \frac{x^2}{v^2} \tag{3.1}$$

This results in a velocity stack in the radon domain with (Yilmaz, 2001):

$$u(v,\tau) = \sum_x d(x,t = \sqrt{\tau^2 + \frac{x^2}{v^2}}) \tag{3.2}$$

and for the transformation back into the x-t domain :

$$P(x,t) = \sum_v u(v,\tau = \sqrt{t^2 + \frac{x^2}{v^2}}) \tag{3.3}$$

Seismic data processing

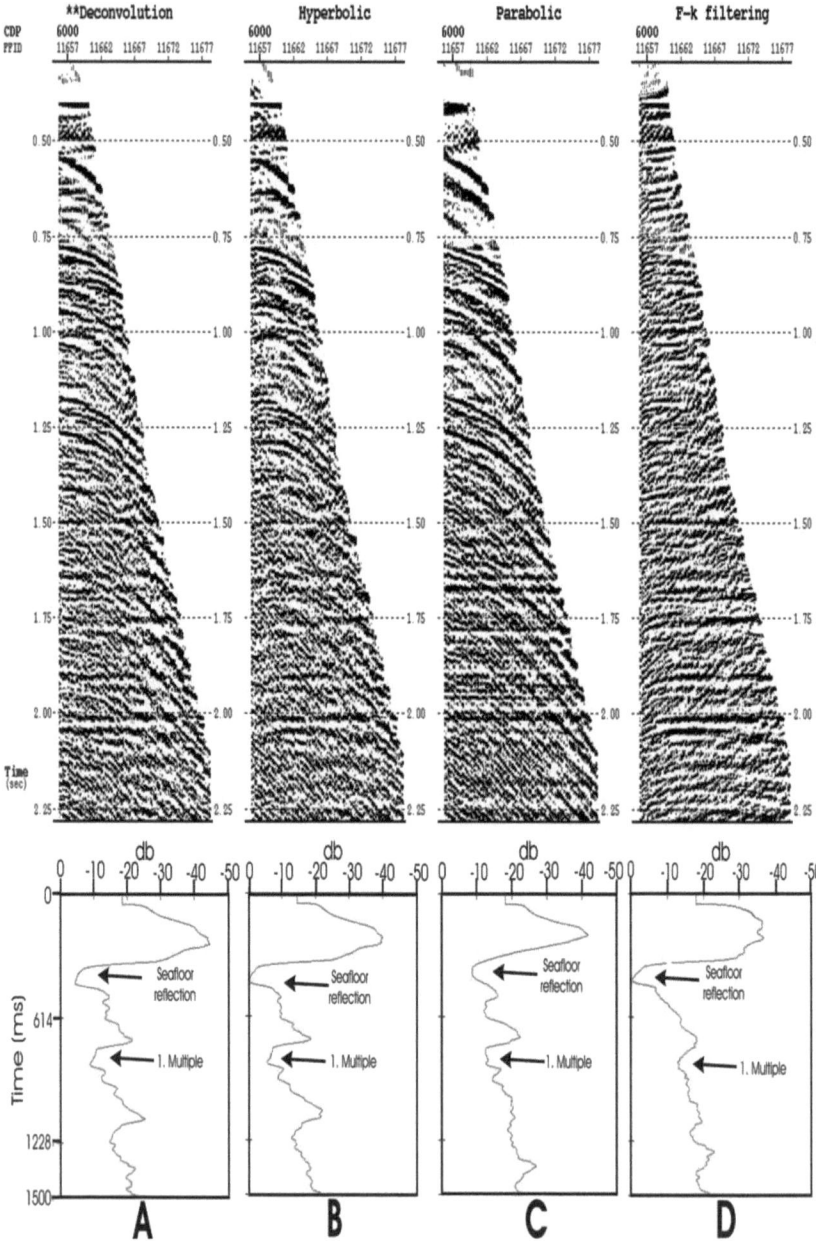

Figure 3.1: CMP gather with four different multiple suppression methods (predictive deconvolution, hyperbolic and parabolic radon transformation, f-k filtering. Corresponding gain analyses of these gathers are shown below.

This transformation does not work ideal for a real dataset. Fundamental problems occur during the transformation in the velocity domain, because of the finite length of the hydrophon cable as well as the discontinuous input function. That means velocities of data with less moveout (small offset) are not resolvable after the transformation and the inversion into the x-t domain do not provide the output signal. To avoid this problem, the hyperbolic radon transformation was developed (Yilmaz, 2001) with the result the time axis (t and τ) will be stretched with the condition: t' = t^2; τ' = τ^2. Hence hyperbolic events in the distance field will be transformed to parabolic events in the radon domain. This transformation will be applied for each point in the Fourier domain using a matrix (least square fit). An inverse Fourier transformation was carried out and the stretching of the time axis was calculated back. Following this, a filter operation in the radon domain and the back transformation of the modelled CDP gather was applied (Sheriff and Geldard, 1995).

Panel b in figure 3.1 shows the result of the hyperbolic radon transformation. The amplitude spectrum below exhibits an attenuation of the multiple of 5 db (from 0 db of the seafloor reflection to -5 db of the first multiple) in comparison to the primary signal. The result looks quite similar to that of the predictive deconvolution, and also the reduction of the multiple energy shows the same trend in the gain analysis. The results are not convincing enough to apply this method to the whole dataset.

parabolic The normal moveout (NMO) correction of the incoming data must be applied before the parabolic velocity filtering is performed. For the calculation the following equation was used for the moveout correction:

$$t_n = \sqrt{t^2 - \frac{x^2}{v_n^2}} \qquad (3.4)$$

Now, resulting moveouts appear nearly parabolic (Yilmaz, 2001):

$$t_n = \tau + q\frac{x^2}{2} \qquad (3.5)$$

The forward transformation describes the segmentation of the NMO-corrected data into a sum of parabolic events. The primary signals differ from multiple signals because of the moveout of the parabolas (Geissler, 2000). The best fit for NMO-corrected CMP gathers is the transfer into parabolas rather than the linear τ-ρ-transformation. The NMO corrected multiples with p-wave velocities of the primary signals describe parabolas in a first approximation (Hampson, 1986). The dip of the summation path can be joined with the velocities. Thereby it is possible to terminate the great number of parabolas applied to a defined dip and velocity interval, during the forward and reverse transformation (Geissler, 2000). Now it was possible to filter out the NMO-corrected primary reflections before the transformation back into the x-t domain. Therefore, only the multiple reflections were transformed back. The result is a seismogram with multiple reflections, which has to be subtracted from the original seismogram.

The result of the parabolic radon transformation is shown in figure 3.1 in panel c. Within the seismic section above at 1.25 s a clear indication of the second seafloor multiple can be observed. It was not successful to suppress multiples of higher order. This result is also reflected in the gain analysis below. The amplitude spectrum below exhibits an attenuation of the multiple of 3 db (from -9 db of the seafloor reflection to -12 db of the first multiple) in comparison to the primary signal. Nearly no attenuation of the first seafloor multiple,

Seismic data processing

in contrast to the seafloor energy could be reached. The stacking result in figure 3.4d demonstrates that the multiples mask completely deeper signals.

c) f-k-filtering

Through a transformation of a seismogram from a x-t to f-k domain it is possible to separate multiple and primary signals, because of their different gradient in the x-t domain. The f-k transformation is defined as:

$$F(f,k) = \int_{-\infty}^{\infty} \int_{-\infty}^{\infty} f(x,t) \cdot e^{-i(2\pi ft + kx)} dx dt \qquad (3.6)$$

A monochromatic wave is displayed on a single point in the f-k domain. On a line trough origin, all frequencies are displayed, that have the same declination in the x-t domain. The coherence between declination in the x-t domain and the declination in the f-k domain will quantified by

$$\frac{dx}{dt} = v = \frac{f}{k} \qquad (3.7)$$

The multiple suppression by f-k filtering (Figure 3.2) was applied with the FOCUS module ZMULT. The data were corrected with the result, that primary signals are over-corrected in the same dimension like the multiples are under-corrected. After this transformation in the f-k domain the primary signals can be observed in the x-t domain in the area of negative wave numbers. The reason is their "negative" dip (increasing travel time with increasing offset). In contrast, the multiples have a "positive" dip (decreasing travel time with decreasing offset) in the x-t domain and are displayed in the area of positive wave numbers. Therefore, the multiples can be separated from the primary signals. The areas of positive and negative wave numbers will be transformed back into the x-t domain after the positive areas is set to 0. The last panel on figure 3.1d images the result of the f-k filtering. The first multiple at 780 ms could be nearly suppressed. The amplitude spectrum below exhibits an attenuation of the multiple of 13 db (from 0 db of the seafloor reflection to -13 db of the first multiple) in comparison to the primary signal. The seismic sections show continuously horizontal reflections, which result after stacking in clearly visible reflectors (see figure 3.5).

3.1.2 Results of the multiple suppression

The results of the applied methods are demonstrated on profile AWI-20030390 recorded with the 3000 m long streamer (Figs. 3.3, 3.4, 3.5). Figure 3.3a shows the original stacked data without any application of multiple suppression methods. In this time section the first water bottom multiple is visible at around 850 ms. On the basis of well imaged primary signals and the almost completely suppressed second water bottom multiple I assume that the velocity model for the shelf region of this profile provides the best subsurface image. Between CDP 5500 and 7000 weak inclined reflectors (seafloor multiples) are visible. Prior to interpretation it is necessary to extract these signals from the primary signals.

The usage of a predictive deconvolution (Fig. 3.3b) reached a reduction of the first multiple, and primary signals between 500 and 1200 ms can be better observed now. The rest of the energy of the first water bottom multiple shows a low-frequency signal in comparison to the signal on figure 3.3a, which represents a high-frequency signal. A complete suppression was not achieved, which can be explained with the violation of the preconditions like a static

3.1 Seismic reflection

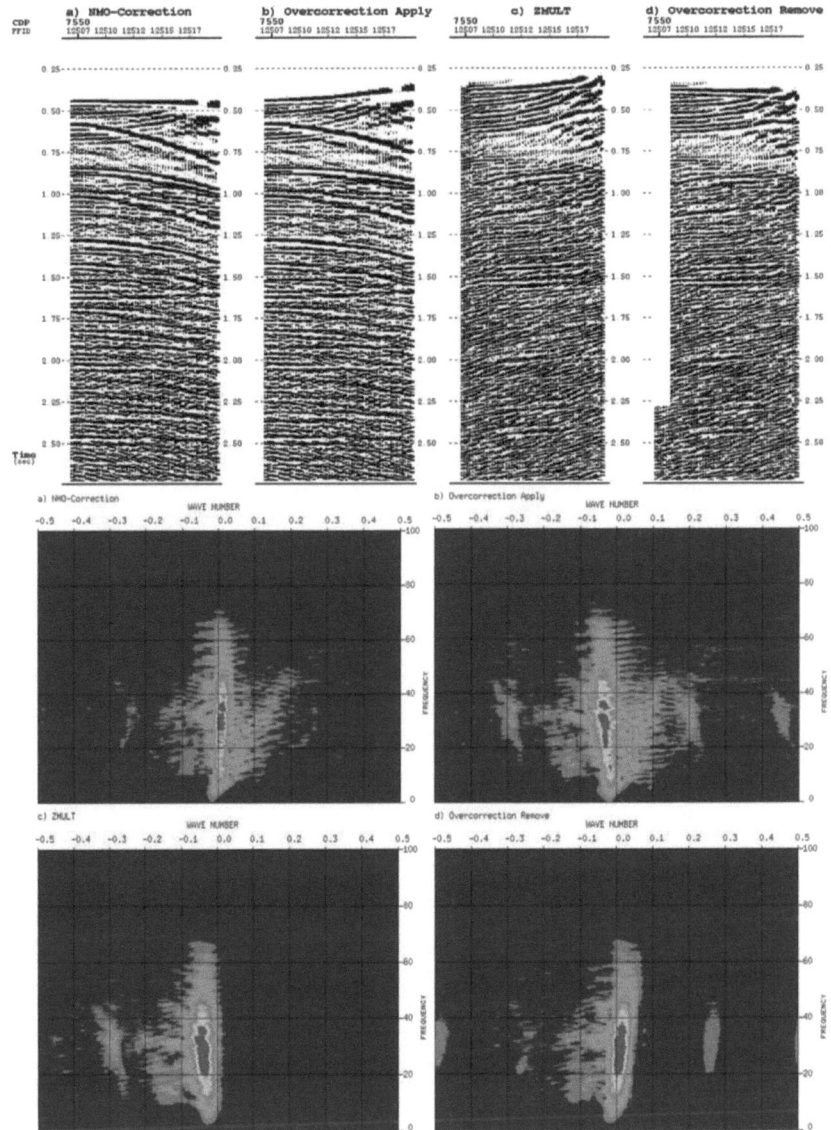

Figure 3.2: Results of a f-k filtering of a data set recorded with a 3000 m long streamer: the upper part of the figure shows a CMP gather after several processing steps. The lower part shows the associated f-k spectrum. a) NMO correction with under corrected multiples, displayed in the positive wave number area of the f-k spectrum. The overcorrection b) produce a separation of primary and multiple signals in the f-k spectrum. c) f-k filtering, d) over correction is removed and the rest of the energy is only visible at the frequency axis; multiple removed CMP gather

source and minimum-phase signal.

Results after a hyperbolic radon transformation (Fig. 3.4c) show no change to the stacked section (3.3a) in contrast to the result of the parabolic radon transformation. With the parabolic radon transformation, the deeper signals (below 1s TWT) were completely masked and the geological structure is no longer visible (Fig. 3.4d). Also multiples of higher order could not be suppressed. Different parameters were tested but a satisfactory result could not be reached with this method. A better result could be achieved with the hyperbolic radon transformation, but an improvement of the stack result could be not obtained.

The best result could be achieved by the f-k filtering (Fig. 3.5e). The parameters for the over-correction of primary signals were good enough to separate these signals from the multiples. Under-corrected signals were removed (Fig. 3.2d) and the remaining over-corrected signals were transfered back for stacking. Therefore, a mute for the near offset (300 m) traces to improve the end result (Fig. 3.2d) was applied. The near offset traces from multiple signals show often no dips. Thus, it is difficult to distinguish between over and under correction.

In general, it was successful to suppress the seafloor multiples in the shelf region by using the 3000 m streamer. It was more difficult to achieve an improvement for the data set north of the Greenland Fracture zone where the data were recorded with the 600 m streamer. These data provide only limited velocity information, because of the small offset. The primary signals could be not separated after a NMO correction, which is the main pre-condition for the used multiple suppression methods.

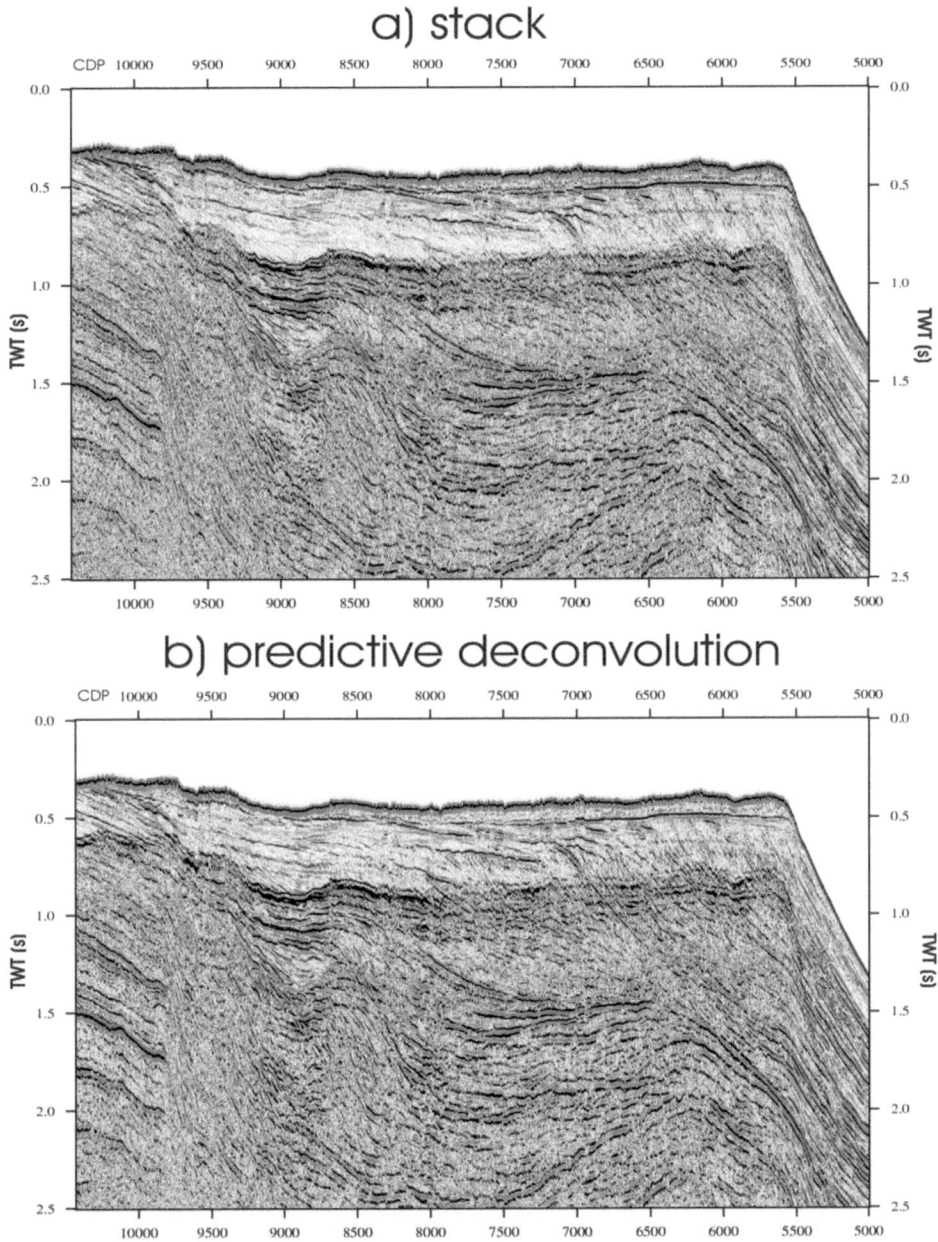

Figure 3.3: Shelf region of profile AWI-20030390. a) stacked section, b) section after applying a predictive deconvolution.

Seismic data processing

c) after hyperbolic radon transformation

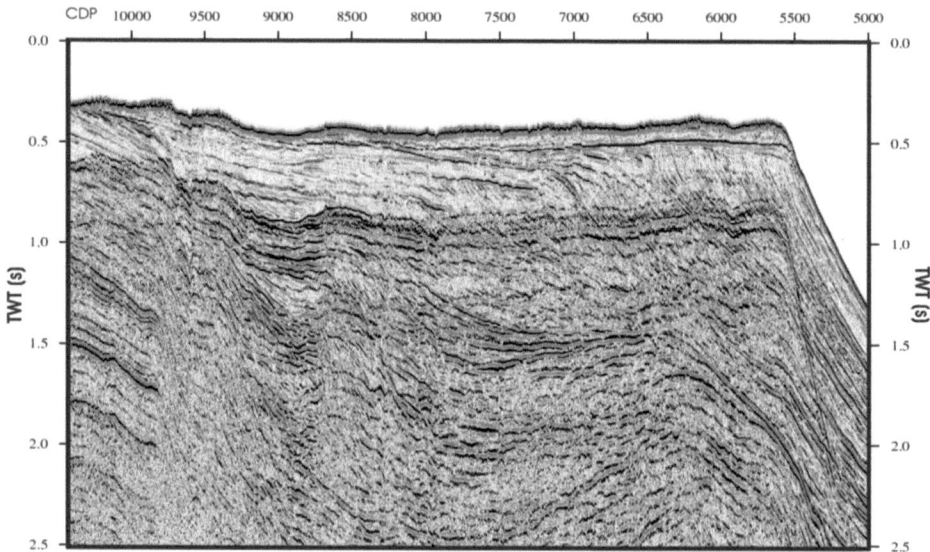

d) after parabolic radon transformation

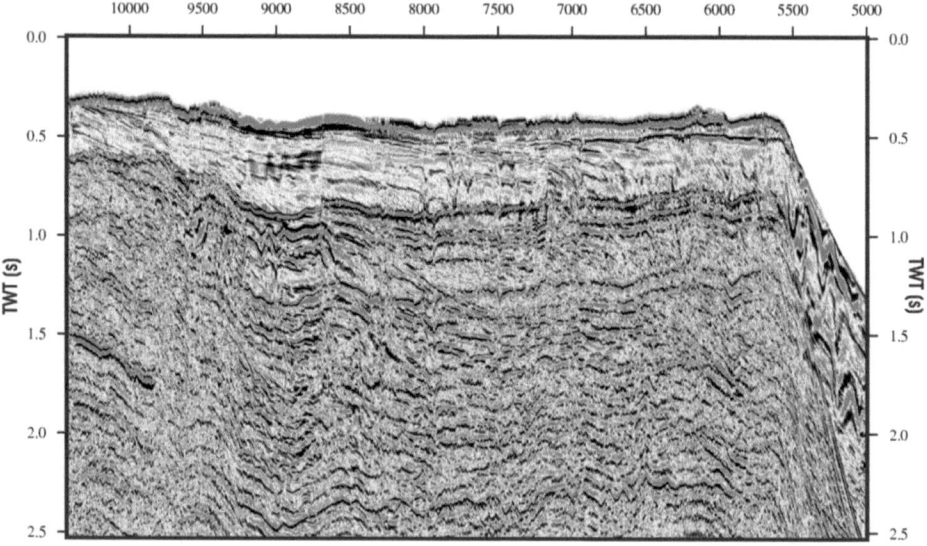

Figure 3.4: Shelf region of profile AWI-20030390. c) section after applying a hyperbolic radon transformation, d) section after applying a parabolic radon transformation.

3.1 Seismic reflection

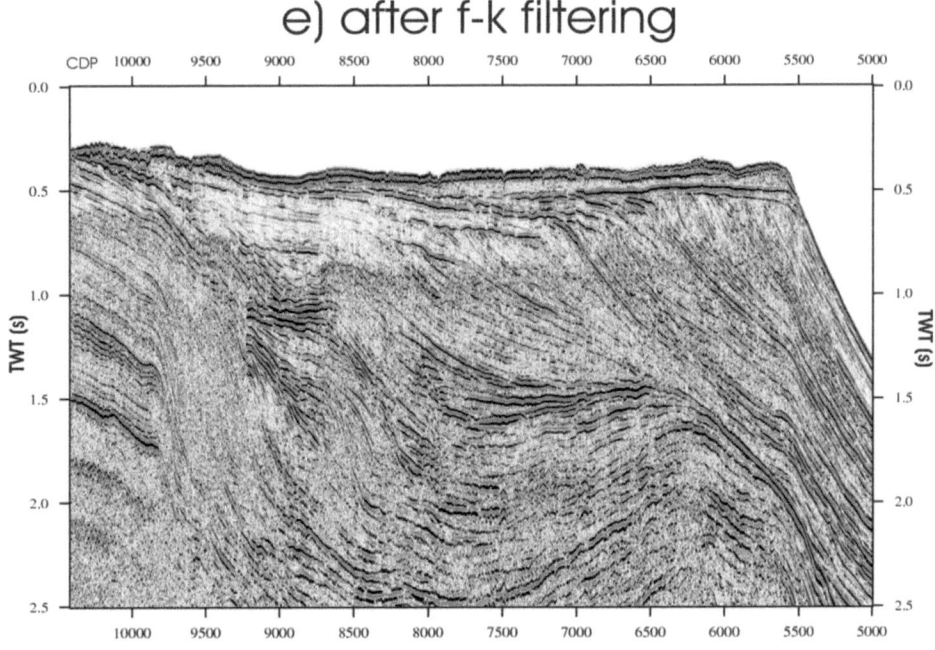

Figure 3.5: Shelf region of profile AWI-20030390. e) section after applying a f-k filtering.

4 References

Anderson, J.B., Kurtz, D.D., Weaver, F.M., (1979). *Sedimentation on the Antarctic continental slope.* In: Doyle, L.J., Pilkey, O. (Eds.), Geology of Continental Slopes. Special Publication-Society of Economic Paleontologists and Mineralogists, vol. 27, 127 pp.

Anderson, J.B., Kurtz, D.D., Domack, E.W., Balshaw, K.M., (1980). *Glacial and glacial marine sediments of the Antarctic continental shelves.* Journal of Geology 88, 399-414.

Anderson, J.B., (1999). *Antarctic Marine Geology.* Cambridge University Press, 289 pp.

Barker, P.F., (1995). *The proximal marine sediment record of Antarctic climate since the late Miocene.* In: Cooper, A.K., Barker, P.F., Brancolini, G., (Eds), Geology and seismic stratigraphy of the Antarctic margin. Antarctic Research Series 68 AGU, Washington, D.C., pp. 25-57.

Brancolini, G., Cooper, A.K., Coren, F., (1995). *Seismic facies and glacial history in the Western Ross Sea Antarctica.* In: Cooper, A.K., Barker, P.F., Brancolini, G. Eds. , Geology and seismic stratigraphy of the Antarctic margin. Antarctic Research Series 68 AGU, Washington, DC, pp. 209-233.

Brozena, J.M., Childers, V.A., Lawver, L.A., Gahagan, L.M., Forsberg, R., Faleide, J.I. and Eldholm, O. (2003). *New aerogeophysical study of the Eurasia Basin and Lomonosov Ridge: Implications for basin development.* Geology, 31, 825-828.

Chalmers, J.A., Pulvertaft, T.C.R., 2001. *Development of the continental margins of the Labrador Sea: a review.* In: R.C.L. Wilson, R.B. Whitmarsh, B. Taylor, N. Froitzheim (Eds.), Non-Volcanic Rifting of Continental Margins: A Comparison of Evidence from Land and Sea, vol. 187, Geological Society of London, London, UK, pp. 77-105.

Channell, J.E.T., Smelror, M., Jansen, E., Higgins, S.M., Lehman, B., Eidvin, T., Solheim, A., (1999). *Age models for glacial fan deposits off East Greenland and Svalbard (Sites 986 and 987).* In: Raymo, M.E., Jansen, E., Blum, P., Herbert, T.D. (Eds.), Proceedings of the Ocean Drilling Program, Scientific Results, vol. 62, pp. 149-166.

Cooper, A.K., Barrett, P.J., Hinz, K., Traube, V., Leitchenkov, G., Stagg, H.M.J., (1991). *Cenozoic prograding sequences of the Antarctic continental margin: a record of glacio-eustatic and tectonic events.* Marine Geology 102, 175-213.

Darby, D. A., Zimmerman, P., (2008). *Ice-rafted detritus events in the Arctic during the last glacial interval, and the timing of the Innuitian and Laurentide ice sheet calving events.* Polar Research, Volume: 27, Number: 2,114-127.

Denton, G. H., Prentice, M. L., and Burckle, L. H., (1991). *Cenozoic history of the Antarctic Ice Sheet.* In: Tingey, R. J. (Eds.) The Geology of Antartica, Clarendon Press, Oxford,

365-433.

De Santis, L., Anderson, J.B., Brancolini, G., Zayatz, I., (1995). *Miocene glacio-marine facies analysis on the central and eastern continental shelf of the Ross Sea, Antarctica*. In: Cooper, A.K., Barker, P.F., Brancolini, G. Eds., Geology and Seismic Stratigraphy of the Antarctic Margin. Antarctic Research Series 68 AGU, Washington, DC, pp. 209-233.

De Santis, L., Brancolini, G., Donda, F., (2003). *Seismic-stratigraphic analysis of the Wilkes Land continental margin (East Antarctica). Influence of glacially-driven processes on the Cenozoic deposition*. Deep-Sea Research. Part 2. Topical Studies in Oceanography 50 (8-9), 1563-1594.

Domack, E.W., (1982). *Sedimentology of glacial and glacial marine deposits on the George VAdelie continental shelf, East Antarctica*. Boreas 11, 79-97.

Dowdeswell, J.A., Kenyon, N.H., Laberg, J.S., (1997). *The glacier-influenced Scoresby Sund Fan, East Greenland continental margin: evidence from GLORIA and 3.5 kHz records*. Marine Geology, 143, 207-221.

Dowdeswell, J.A., Ó Cofaigh, C., and Pudsey, C.J., (2004). *Thickness and extent of the subglacial till layer beneath an Antarctic paleo-ice stream*. Geology, v. 32, p. 13-16.

Ehlers, B.-M. and Jokat, W., (2008). *Analysis of subsidence in crustal roughness for ultra-slow spreading ridges in the northern North Atlantic and the Arctic Ocean*. Geophysical Journal International, Vol. 177, Issue 2, pp. 451-462.

Ehlers, B.-M. and Jokat, W., (submitted 2008a). *Palaeobathymetric study of the North Atlantic and the Eurasia Basin*. Geophysical Journal International.

Ehlers, B.-M., Butzin, M., Grosfeld, K., Jokat, W., (submitted 2008b). *A palaeoceanographic modelling study of the Cenozoic northern North Atlantic and the Arctic Ocean*. Global and Planetary Change.

Eidvin, T., Brekke, H., Riis, F., Renshaw, D.K., (1998). *Cenozoic stratigraphy of the Norwegian Sea continental shelf, $64°N68°N$*. Norsk Geologisk Tidsskrift 78, 125-152.

Einsele, G., (1992). *Sedimentary Basins: Evolution, Facies, and Sediment Budget*. Springer Verlag, Berlin Heidelberg, New York.

Eldholm, Olav; Windisch, Charles C., (1974). *Sediment Distribution in the Norwegian-Greenland Sea*. Geological Society of America Bulletin, Volume: 85, Number: 11, 1661-1676.

Eldrett, J.S., Harding, I.C., Wilson, P.A., Butler, E. & Roberts, A.P., (2007). *Continental ice in Greenland during the Eocene and Oligocene*. Nature, 446, 176-179.

References

Engen, Ø, Faleide, J. I., Dyreng, T. K. (2008). *Opening of the Fram Strait gateway: A review of plate constraints*. Tectonophysics, 450, 51-69.

Escutia, C., Warnke, D.A., Acton, G.D., Barcena, A., Burckle, L., Canals, M., Frazee, C.S., (2003). *Sediment distribution and sedimentary processes across the Antarctic Wilkes Land margin during the Quaternary*. Deep-Sea Research. Part 2. Topical Studies in Oceanography 50, 1481-1508.

Escutia, C., De Santis, L., Donda, F., Dunbarc,R.B., Cooper, A.K., Brancolinib, G., Eittreim, S.L., (2005). *Cenozoic ice sheet history from East Antarctic Wilkes Land continental margin sediments*, Global and Planetary Change 45, 51-81.

Forsberg, C.F., Solheim, A., Elverhøi, A., Jansen, E., Channell, J.E.T., Andersen, E.S., (1999). *The depositional environment of the western Svalbard margin during the Late Pliocene and the Pleistocene: sedimentary facies changes at site 986*. In: Raymo, M.E., Jansen, E., Blum, P., Herbert, T.D. (Eds.), Proceedings of the Ocean Drilling Program Scientific Results, vol. 162.

Geissler, W., (2000). *Marine seismische Untersuchungen am nördlichen Kontinentalrand von Spitzbergen*, Diploma thesis, Technical university, Freiberg.

Gradstein, F., Ogg, J. and Smith, A. (eds.), (2004). *A geological time scale 2004*. Cambridge University Press, Cambridge, United Kingdom (GBR).

Hamann, N.E., Whittaker, R.C. & Stemmerik, L., (2005). *Geological development of the Northeast Greenland Shelf*. In: Proceedings of the 6th Petroleum Geology conference Petroleum Geology: North-West Europe and Global Perspectives, eds Doré, A.G. & Vining, B.A., Geological Society, London, pp. 887-902.

Hambrey, M.J., Larsen, B., Ehrmann, W.U., (1989). *Forty million years of Antarctic glacial history yielded by Leg 119 of the Ocean Drilling Program*. Polar Record, vol. 25, no. 153, pp. 99-106.

Hampson, G., (1986). *The relationship of pre-stack apparent velocity filtering to the symmetry of the CMP stack response*. First Break, Volume: 5, Number: 10, 359-377.

Helland, P.E. & Holms, M.A., (1997). *Surface textural analysis of quartz sand grains from ODP Site 918 off the southeast coast of Greenland suggests glaciation of southern Greenland at 11 Ma*. Palaeogeogr. Palaeoclimatol. Palaeoecol., 135, 109-121.

Hesse, R., Klauke, I., Ryan, W.B.F., Piper, D.J.W., (1997). *Ice-sheet sourced juxtaposed turbidite systems in Labrador Sea*. Geoscience Canada 24 (No.1), 3-12.

Hinz, K., Mutter, J.C., Zehnder, C.M. and Group, N.S., (1987). *Symmetric conjugation of continent-ocean boundary structures along the Norwegian and East Greenland margins*.

Mar. Petrol. Geol., 3, 166187.

Hinz, K., Meyer, H. and Miller, H., (1991). *North-east Greenland shelf north of 79° N: results of a reflection seismic experiment in sea ice.* Marine and Petroleum Geology, Vol. 8, pp. 461-467.

Hjelstuen, B. O., Eldholm, O., Faleide, J. I., (2007). *Recurrent Pleistocene mega-failures on the SW Barents Sea margin.* Earth and Planetary Science Letters, Volume: 258, Number: 3-4, 605-618.

Jakobsson, M. et al., (2007). *The early Miocene onset of a ventilated circulation regime in the Arctic Ocean.* Nature, Vol. 447, 986-990.

Jansen, E., Bleil, U., Henrich, R., Kringstad, L., Slettemark, B., (1988). *Paleoenvironmental changes in the Norwegian Sea and the northeast Atlantic during the last 2.8 m.y.: deep sea drilling project/ocean drilling program sites 610, 642, 643 and 644.* Paleoceanography 3, 563-581.

Jansen, E., Fronval, T., Rack, F., Channell, J.E.T., (2000). *Pliocene- Pleistocene ice rafting history and cyclicity in the Nordic Seas during the last 3.5 Myr.* Paleoceanography 15, 709-721.

Knies, J. and Gaina, C., (2008). *Middle Miocene ice sheet expansion in the Arctic: Views from the Barents Sea.* Geochemistry Geophysics Geosystems, 9, Q02015, doi:10.1029/2007GC001824.

Kuvaas, B., Kristoffersen, Y., (1991). *The crary fan: a trough-mouth fan on the Weddell sea continental margin, Antarctica.* Marine Geology 97, 345-362.

Kuvaas, B., Kristoffersen, Y.K., Leitchenkov, G., Guseva, J., Gandjukhin, V., (2004). *Seismic expression of glaciomarine deposits in the eastern Riiser-Larsen Sea, Antarctica.* Marine Geology 207, 1-15.

Laberg, J.S., Vorren, T.O., (1996). *The Middle and Late Pleistocene evolution of the Bear Island Trough Mouth Fan.* Global and Planetary Change 12, 309-330.

Laberg et al., (2005). *Cenozoic alongslope processes and sedimentation on the NW European Atlantic margin.* Marine and Petroleum Geology, Volume: 22, Number: 9-10, 1069-1088.

Larsen, H.C., (1990). *The East Greenland Shelf.* In: Grantz, A., Johnson, L., Sweeney, J.F. (Eds.), The Arctic Ocean region. The geology of North America L, vol. 185-210. Geological Society of America, Boulder, CO, pp. 185-210.

Larsen, H.C., Saunders, A.D., Clift, P.D., Beget, J., Wei, W., Spezzaferri, S., (1994). *ODP Leg 152 scientific party, seven million years of glaciation in Greenland.* Science 264, 952-955.

References

Lawver, L.A., Gahagan, L.M., (1998). *Opening of Drake Passage and its impact on Cenozoic ocean Circulation.* In: Crowley TJ, Burke KC (eds) Tectonic boundary conditions for climate reconstructions. Oxford University Press, Oxford, pp 212-223.

Lawver, L.A., Gahagan, L.M., (2003). *Evolution of Cenozoic seaways in the circum-Antarctic region.* Paleogeogr Palaeoclim Palaeoecol 198, 11-37.

Leinweber, V.T. (2006). *Abschätzung von Sedimentmächtigkeiten in der Framstraße anhand magnetischer Daten.* Diploma thesis, Karl-Franzens university, Graz.

Militzer, H. & Weber, F., (1987). *Angewandte Geophysik 3, Seismik.* Springer-Verlag, Akademie-Verlag Berlin.

Miller, H., Henriet, J.P., Kaul, N. and Moons, A., (1990). *A fine-scale stratigraphy of the eastern margin of the Weddell Sea.* In: Bleil U. and Thiede J. (eds.) Geological history of the Polar Oceans: Arctic versus Antarctic, pp 131-161. Kluwer Academic Publishers.

Moore, G.T., Sloan, L.C., Hayashida,D.N. and Umrigar,N.P., (1992). *Paleoclimate of the Kimmeridgian/Tithonian (Late Jurassic) world: II. Sensivity tests comparing three different paleotopographic settings.* Paleogeogr. Paleoclimatol. Paleoecol., 95, 229-252.

Myhre, A.M., Thiede, J., Firth, J.V., Shipboard Scientific Party, (1995). *North Atlantic-Arctic Gateways.* Proc. ODP Init. Rep., 151.

Nielsen, T., De Santis, L., Dahlgren, K.I.T., Kuijpers, A., Laberg, J.S., Nygård, A., Praeg, D., Stoker, M.S., (2005). *A comparison of the NW European glaciated margin with other glaciated margins.* Marine and Petroleum Geology, Vol. 22, no. 9-10, pp.1149-1183.

Rise,L., Ottesen, D., Berg, K., Lundin, E., (2005). *Large-scale development of the mid-Norwegian margin during the last 3 million years.* Marine and Petroleum Geology 22, 33-44.

Sættem, J., Poole, D.A.R., Ellingsen, L., Sejrup, H.P., (1992). *Glacial geology of outer Bjrnyrenna, southwestern Barents Sea.* Marine Geology 103, 15-51.

Schlindwein, V. and Jokat, W., (1999). *Structure and evolution of the continental crust of northern east Greenland from integrated geophysical studies.* Journal of Geophysical Research 104(B7), 15227-15245.

Schlüter, P. and Uenzelmann-Neben, G., (2007). *Seismostratigraphic analysis of the Transkei Basin: A history of deep sea current controlled sedimentation.* Marine Geology, 240(1-4), 99-111.

Shackleton, N.J., Imbrie, J., Pisias, N.G., (1988). *The evolution of oceanic oxygen-isotope variability in North Atlantic over the past three million years.* Philosophical Transactions of the Royal Society of London, Series B: Biological Sciences, vol. 318, no. 1191, pp. 679-688.

References

Sheriff, R.E. und Geldardt, L.P., (1995). *Exploration Seismology*, Second Edition, USA.

Shipboard Scientific Party, (1989). *Site 739*. Proc. ODP, Init. Rep., 119: 289-344.

Skogseid, J., Planke, S., Faleide, J.I., Pedersen, T., Eldholm, O. and Neverdal, F., (2000). *NE Atlantic Continental rifting and volcanic margin formation*. In Dynamics of the Norwegian Margin, Vol. 167, pp. 295-326.

Stemmerik, L., (1993). *High frequency sequence stratigraphy of a siliciclastic influenced carbonate platform, lower Moscovian, Amdrup Land, North Greenland*. Geol. Soc. Spec. Publ., 104, 347-365.

Stoker, M.S. et al. (2005). *Sedimentary and oceanographic responses to early Neogene compression on the NW European margin*. Marine and Petroleum Geology 22, 1031-1044.

Talwani, M. und Eldholm, O., (1977). *Evolution of the Norwegian-Grenland-Sea*. Geological Society of America Bulletin 88, 969-999.

Tsikalas, F., Faleide, J.I., Eldholm, O. & Wilson, J., (2005). *Late Mesozoic- Cenozoic structural and Stratigraphic correlations between the conjugate mid-Norway and NE Greenland continental Margins, Geological development of the Northeast Greenland Shelf*. In: Proceedings of the 6th Petroleum Geology Conference Petroleum Geology: North-West Europe and Global Perspectives, eds Doré, A.G. & Vining, B.A., Geological Society, London, pp. 785-801.

Tucholke, B. E.; Sawyer, D. S.; Sibuet, J. C., (2007). *Breakup of the Newfoundland-Iberia Rift*. Geological Society Special Publications, Volume: 282, 9-46.

Verhoef, J., Macnab, R., Roest, W., Arjani-Hamed, J. and the project team, (1996). *Magnetic anomalies of the Arctic and the North Atlantic and adjacent land areas*. GAMMA5 (Gridded Aeromagnetic and Marine Magnetics of the North Atlantic and Arctic, 5km). Geological Survey of Canada, Open File 3125a (CD-Rom).

Voss, M. and Jokat, W., (2007). *Continent-ocean transition and voluminous magmatic underplating derived from P-wave velocity modelling of the East Greenland continental margin*. Geophysical Journal International, Volume 170, Issue 2, Page 580-604.

Wagner, T., Hölemann, J.A., (1995). *Deposition of organic matter in the Norwegian-Greenland sea during the past 2.7 million years*. Quaternary research 44, 355-366.

Weigel et al., (1995). *Investigations of the East Greenland Continental Margin between 70° and 72°N by Deep Seismic Sounding and Gravity Studies*. Marine geophys. Researches 17, 167-199.

Winkler, A., Wolf-Welling, T.C.W., Stattegger, K. & Thiede, J., (2002). *Clay mineral sedimentation in high northern latitude deep-sea basins since the Middle Miocene (ODP Leg*

151, NAAG). Int. J. Earth Sci., 91, 133-148.

Wong, H.K., Lüdmann, T., Baranov, B.V., Karp, B.Ya., Konerding, P., Ion, G., (2003). *Bottom current-controlled sedimentation and mass wasting in the northwestern Sea of Okhotsk*. Marine Geology 201, 287-305.

Wright, J.D., Miller, K.G., Fairbanks, R.G., (1992). *Early and Middle Miocene Stable Isotopes: Implications for Deep water Circulation and Climate*. Paleoceanography, 7(3), 357-389.

Yilmaz, Ö., (2001). *Seismic Data Analysis, Processing, Inversion and Interpretation of Seismic Data*. Society of Exploration Geophysics, SEG.

Zachos, J., Pagani, M., Sloan, L., Thomas, E., Billups, K., (2001). *Trends, Rhythms, and Aberrations in Global Climate 65 Ma to Present*. Science 292, 686-693.

Ziegler, P.A., (1990). *Geological Atlas of Western and Central Europe*. Geological Society Publishing House, London, U.K., 239 pp.

References

5 Paper 1

A seismic study along the East Greenland margin from 72°N 77°N

Daniela Berger and Wilfried Jokat

Geophysical Journal International (2008), Vol. 174, Issue 2, pp. 743-754.

accepted 2008 March 17. Received 2007 December 4; in original form 2007 June 21

5.1 Abstract

Around 4370 km of new seismic reflection data, collected along the East Greenland margin between 71°30N and 77°N in 2003, provide a first detailed view of the sediment distribution and tectonic features along the East Greenland margin. After processing and converting the data to depth, we correlated ODP-Site 913 stratigraphy into the new seismic network. Unit GB-2 shows the greatest glacial sediment deposits beneath the East Greenland continental shelf. This unit is characterized by the beginning of prograding sequences and has according to our stratigraphic correlation a Middle Miocene age. It might have been caused by rapid changes in sea level and/or glacial erosion by an early ice sheet or glaciers along the coast. A basement high, presumably a 360 km long basement structure at 77°N - 74°54N prevents continuous sediment transport from the shelf into the deep sea area in times before 15 Myrs. The origin of this prominent structure remains speculative, since no rock sample from this structure is available.
Seaward dipping reflectors at the eastern flank of this structure strongly support that it is a volcanic construction and most likely emplaced on continental or transitional crust. The compilation of sediment thickness provide an insight into the regional sediment distribution in the Greenland Basin. An average sediment thickness of 1 km is observed. The north bordering Boreas Basin has a sediment thickness of 1.8 km close to the GFZ.

Key words: Controlled source seismology; Continental margins: divergent; Arctic region.

5.2 Introduction

During Permian and Tertiary times, rift systems separated Pangaea into North America, Eurasia, and Gondwana (Moore et al., 1992a). The late Permian to Middle Jurassic initial phase of Pangaea breakup invoked the southward and westward propagation of the proto-Norwegian-Greenland Sea (NGS) and the Tethys rift systems, whereas the late Jurassic to Paleogene break-up phase was dominated by the stepwise northward propagation of the central Atlantic (Ziegler, 1990). The rapid opening of the central Atlantic took place during the late Jurassic and Early Cretaceous, and intensified rifting in the proto-NGS (Ziegler, 1990). The tectonic evolution of the conjugate Greenland and Norwegian margins has faced compression, extension, magmatism and subsidence phases since the Devonian collapse of the Caledonian orogene and until early Eocene continental breakup at 56 Ma (Talwani and Eldholm, 1977; Skogseid et al. 2000). The Cenozoic geological history can be interpreted from the marine sediments and the crustal structure of the East Greenland margin. First seismic investigations in the Greenland Sea were performed already before 1971 (Ewing and Ewing, 1959; Eldholm and Windisch, 1974). Only a few seismic refraction tracks are located along the Greenland continental margin. The data show a basement high between 75.3°N, 10.5°W and 76.4°N, 5°W, but the landward extension of this feature was not imaged (Eldholm and Windisch, 1974). Ostenso and Wold (1971) published an aeromagnetic profile across the feature that shows an abrupt decrease in anomaly amplitude, suggesting the presence of a magnetic quiet zone. Hinz et al. (1987) gave the first detailed insight into the sedimentary and tectonic structure of the northern East Greenland shelf, based on a seismic reflection profile recorded in 1981. This profile is located almost at the same position like profile 20030390 (Fig. 1), and covers the deep sea part of the Greenland Basin. However, it terminates just close to the shelf edge. "Seaward dipping reflectors" (SDRs) beneath the East Greenland continental slope were recognized on a structural high of which the dipping events create the seaward flank (Hinz et al., 1987). Within the basement structure a basement scarp was identified, the Greenland Escarpment, and interpreted as the landward limit of the structural high.

Mutter and Zehnder (1988) introduce results from two-ship Expanded Spread and Wide Aperture CDP Profiling (ESP) measurements off the East Greenland margin. These measurements were made to investigate the deep crustal structure during the onset of seafloor spreading. Seaward of a region of low velocity crust SDRs occur within thick crust interpreted to be a product of voluminous melt production enhanced by convective partial melting processes (Mutter and Zehnder, 1988). A structural high east of the Greenland Escarpment shows at the top a seismic velocity of 4 km/s. Existing studies show a volcanic influenced region south of 76°N.

The seismic database remained extremely poor across the margin, since the sea ice prevented any systematic data acquisition. Another seismic experiment (KANUMAS), carried out in the years 1991 - 1995 gathered multichannel seismic data along the East Greenland shelf between 72°N and 79°N. In total, KANUMAS acquired 6839 km of multichannel seismic data (Hamann et al., 2005) mostly across the inner shelf. In spite of these geophysical investigations, geological structure of this part of the world is still relatively poorly known. Based on the same dataset Tsikalas et al. (2005) tried to make a first correlation with the KANUMAS data, based on reflection character, regional considerations and onshore Greenland geology (Stemmerik, 1993). They divided the sedimentary column into two units, Plio/Pleistoce and Tertiary sediments. In their modell all prograding sediments at 76°N are of Plio/Pleistocene age.

Magnetic data from the Geological Survey of Norway (NGU) (Verhoef et al., 1996) provided a first information for the location of the continent-ocean boundary in the Greenland Basin, based on the identification of seafloor spreading anomalies. Other markers typical of continent-ocean transition zones are volcanic SDRs, which have been identified at the conjugate margins off Norway (Hinz et al., 1987, Tsikalas et al. 2005), and off southeast Greenland (Planke and Alvestad, 1999). Recent deep seismic sounding data off East Greenland provided good constraints for the structure and width of the continent-ocean transition (Voss and Jokat, 2007). It seems to be located more or less underneath the present shelf edge.

A little more is known about the glacial sediments. The East Greenland shelf might have been influenced since the middle Late Miocene ($\sim$7 Ma; Larsen et al., 1994) by advancing and retreating glaciers/ice streams as documented in progradational and aggradational

5.2 Introduction

sequences of glacio-marine sediments (Vanneste et al., 1995). A large glacio-marine fan system exists off Scoresby Sund at 70°N (O'Cofaigh et al., 2001). Further to the north, the margin exhibits a series of large submarine fans and glacial troughs at 72°N, 73°N and 74°N (Fig. 6.1). These are locations where rapid ice streams flowed, and therefore, where the highest flux of glacial debris across the continental margin, can be expected (Dowdeswell, 1996). The existence of Northern Hemisphere ice sheets can be demonstrated back to the

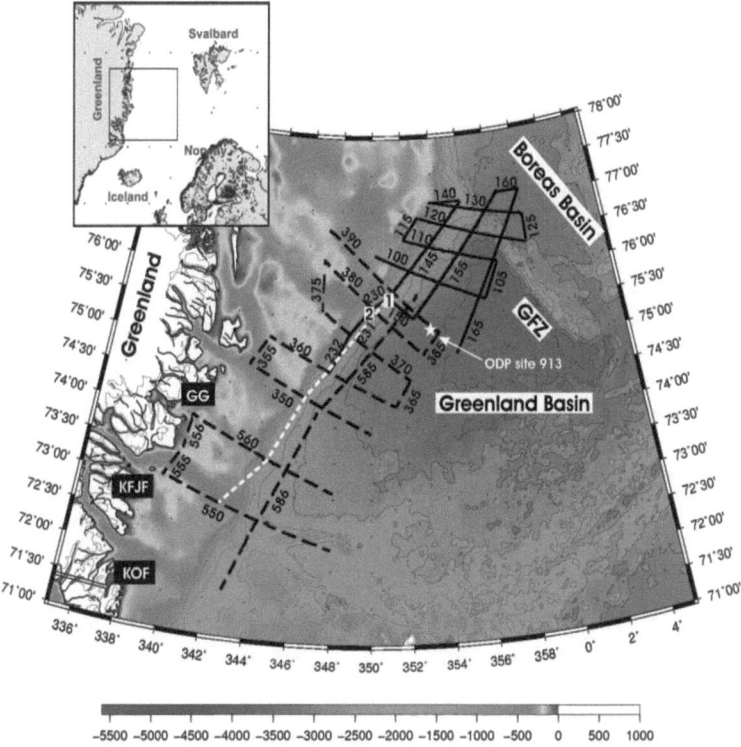

Figure 5.1: Location of the study area with bathymetric contours, and the multichannel seismic (MCS) network recorded with RV "Polarstern" in 2003. The black profiles were acquired with the 600 m long streamer, and the dashed black profiles were acquired with the 3000 m long streamer. ODP drill site 913 is marked with the white star 17 km away from profile AWI-20030390. Numbers 1 and 2 show the location of the sonobuoys 19 (1) and 20 (2). The thin dotted white line along the East Greenland shelf represent the boundary of the ocean-transition zone published by Voss and Jokat (2007). GFZ = Greenland Fracture Zone, GG = Godthåb Gulf, KFJF = Kejser Franz Joseph Fjord, KOF = Kong Oscar Fjord.

middle Miocene, on the basis of proxy data and ice-rafted debris (IRD) from the Nordic seas (Winkler et al., 2002; Helland and Holms. 1997). Eldrett et al. (2007) actually suggest the existence of (at least) isolated glaciers on Greenland between 30 and 38 million years ago based on results of site 913. However, the early glaciation history of the Northern Hemisphere is a subject of controversy. Only one drill hole exists in the Greenland Basin. Ocean Drilling Program (ODP) hole 913 is located in the northwestern Greenland Basin

(Fig. 6.1), where it provides information on the sediment composition and ages (Myhre et al., 1995). On the basis of the drilling results, those authors introduced seven different lithological units. The oldest unit has a middle Eocene age (674.1 - 770.3 m.b.s.f.) and consists of laminated and massive silty clay (Myhre et al., 1995). Basement material could not be recovered.
To advance in the understanding in the tectonic and glacial history of this margin seismic reflection data were collected with RV "Polarstern" in the NGS during the expedition ARK XIX/4 (Jokat et al., 2004). The new seismic data in the NGS comprise 15 profiles recorded with a 600 m long streamer (2309 km), and 13 profiles recorded with a 3000 m long streamer (2062 km). The profiles were arranged parallel and perpendicular to the shelf break (Fig. 6.1), with a line spacing of approximately 50 km. This contribution describes the results of the seismic reflection data interpretation in the Greenland Basin between 71°30N and 77°30N, and provides new insight into the sedimentary structure, glacial history and tectonic evolution of prominent geological formations of this area.

5.3 Seismic data acquisition/processing and mapping procedure

The multichannel survey between 72°N and 78°N was carried out with a cluster of 8 VLF -Guns x 3l (used for measurements with the 600 m streamer) and 5 G -Guns x 8.5l (used for measurements with the 3000 m streamer). Heavy ice in summer 2003 north of 77°N forced the use of the 600 m streamer with 96 channels (Fig. 6.1, black lines), and a hydrophone group spacing of 6.25 m. A 3000 m streamer was used in the south (Fig. 6.1, black dashed lines) with 240 channels and a hydrophone group spacing of 12.5 m. The shot interval on these configurations was 15 seconds, which resulted in a shot distance of about 35 - 40 m. In summer 2004, three seismic reflection profiles and two sonobuoys were additionally recorded along the East Greenland margin to complement the network.
We processed the multichannel seismic data using standard methods. After demultiplexing and applying a spherical divergence correction, the data were sorted and binned into 25 m spaced CDPs. We frequency filtered the data before stacking (15 - 80 Hz for the 600 m long streamer and 10 - 100 Hz for the 3000 m long streamer). For the 3000 m streamer data also a f-k-filtering was applied to suppress the water bottom multiples on the shelf region.
The short streamer provided only limited velocity information for depth conversion, because of the small offset compared to the water depth (>1500 m). Representative velocities for this region are derived from the 3000 m streamer data. The velocity model for profile 20030390 is imaged in figure 6.2. The specified velocities are interval velocities determined by seismic velocity analysis and range between 1.5 and 2.3 km/s for the upper part of the shelf region (CDP: 7350 - 10400). For depths greater than 2 km velocities of 3.2 and 4.3 km/s were used from seismic refraction measurements (Voss and Jokat, 2007). The velocities of deeper reflections were checked by ocean-bottom seismometer deployed along four of the west-east seismic reflection profiles (Voss and Jokat, 2007), and sonobuoy data (Jokat et al., 2005). Acoustic basement velocities were generally extracted from deep sounding data (Voss and Jokat, 2007; Voss and Jokat, submitted 2008). These detailed velocity information is the base for the depth conversion. For profiles north of profile 20030390 we have interpolated the interval velocities of this line to determine an uniform seismic depth section for the entire network.
The greatest inaccuracies are given by the interpolation of the interval velocities along the short streamer profiles in the northern Greenland Basin. In generell, for the interval velocities

5.4 Profile description

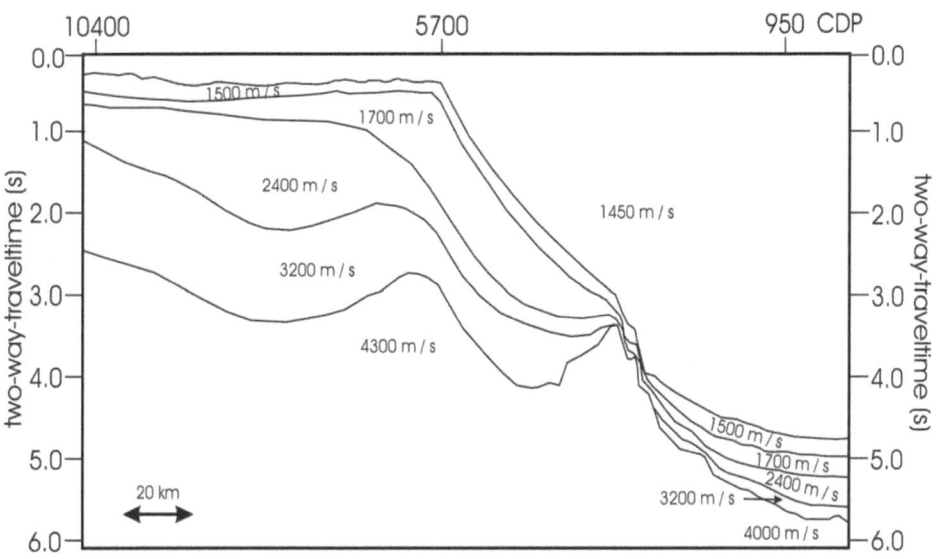

Figure 5.2: Velocity model of profile AWI-20030390 used for depth conversion.

we compute a uncertainty of 150 m/s for velocities and that means for a prominent reflector at profile 20030390 in a borehole (CDP 1252) depth of 360 m.b.s.f. an inaccuracy of $\pm$ 45 m. To create sediment thickness maps, the data were gridded along the seismic reflection profiles and extrapolated to a distance of 3 km each side of the lines using the surface utility from the GMT v4.2.0 package.

5.4 Profile description

Because of different used streamer lengths and different structural information of the seismic profiles we divide our description into two parts. The northern Greenland Basin was investigated using the short streamer. Here, only the slope and abyssal plains were mapped. Additionally, in the southern Greenland Basin the shelf region could be also investigated. Some lines in the southern part of the basin were acquired to the present coast line.

5.4.1 Northern Greenland Basin

Starting in the north, profile 20030130 is located north of the GFZ and profile 20030120 is located at the junction of the Greenland Fracture zone (GFZ) with the East Greenland shelf (Fig. 6.1). On 20030120 (Figs. 6.3, 6.4), a gap between the GFZ and a basement high beneath the shelf is clearly visible in contrast to the northernmost profile. On both profiles seaward of the GFZ rough basement is imaged. A basement high is located at the continental slope, and despite of different water depths (1450 m to 2350 m) and varying size and extend (12.5 km to 20 km) these basement highs look very similar from one line to another. Thick sediment depositions are situated west of the basement highs in the slope

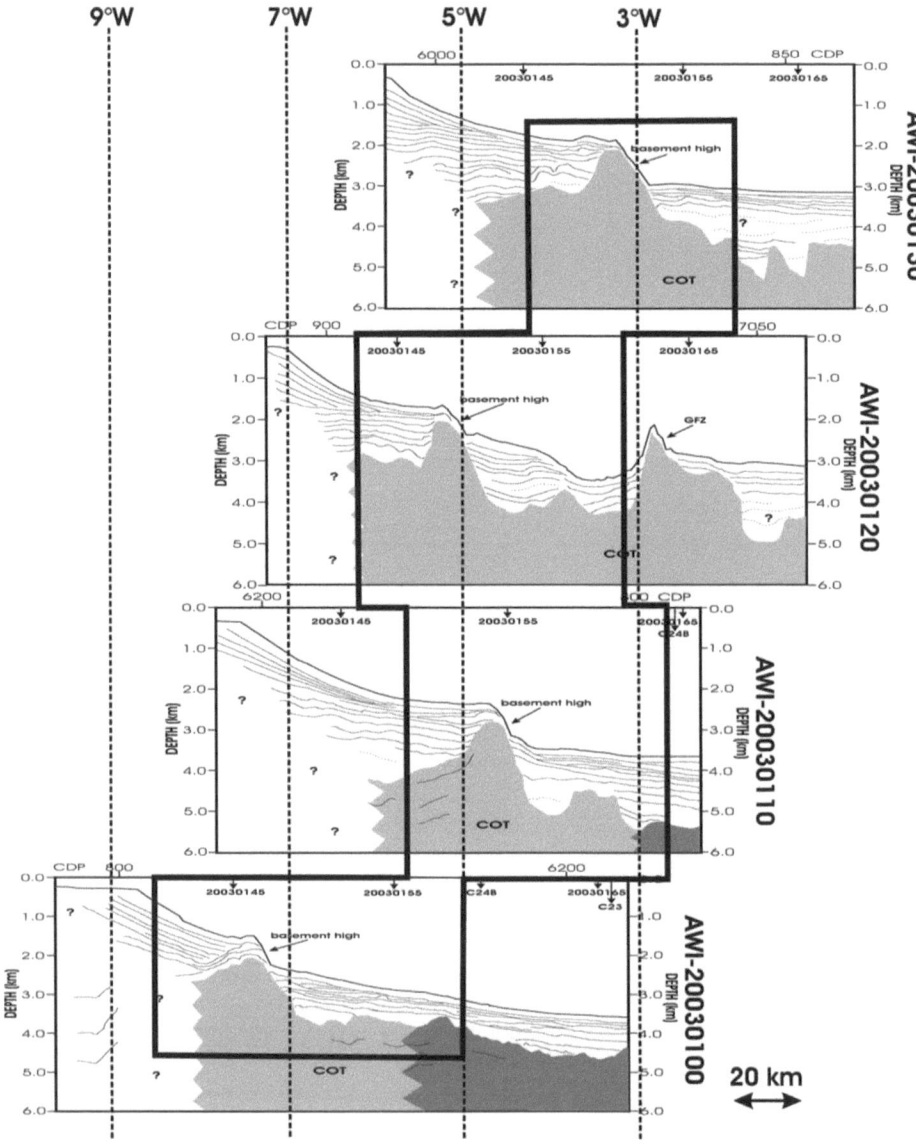

Figure 5.3: Line drawings of profiles AWI-20030130, AWI-20030120, AWI-20030110 and AWI-20030100 (northern Greenland Basin) arranged from north to south and orientated on longitudes 3°W up to 9°W. The numbers at the black arrows represent the ties with crossing profiles at this position. The location of the magnetic spreading anomalies is marked (C24B = 56 Myrs., C23 = 51 Myrs., Gradstein et al., 2004). The heavy black box indicates parts of profiles shown in figure 6.4. The basement is divided into oceanic basement (dark gray) and continent-ocean transition zone (COT, light gray). GFZ = Greenland Fracture Zone.

5.4 Profile description

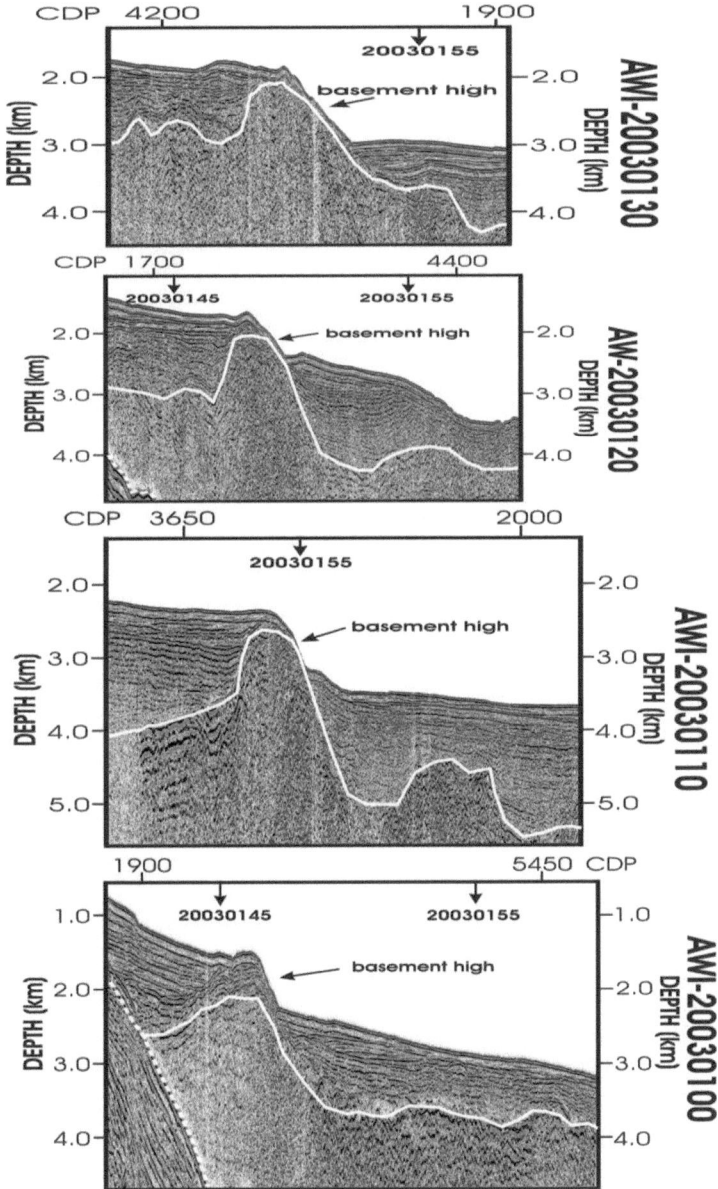

Figure 5.4: Parts of the seismic depth sections of profiles AWI-20030130, AWI-20030120, AWI-20030110 and AWI-20030100. The dashed white line illustrates the water bottom multiple reflection and the white line represents the top of the acoustic basement. The numbers at the black arrows represent the crossing profiles at this position.

region. This structure acted like a wall for the sediment transport into the deep sea. These profiles provide no information on the depth and structure of the acoustic basement beneath the shelf, since the water bottom multiples mask any deeper signals.

5.4.2 Southern Greenland Basin

The next profile (Figs. 6.5, 6.6; 20030390) recorded with the 3000 m streamer is nearly identical with the profile of Hinz et al. (1987) across the East Greenland continental margin, and has a total length of 258 km. ODP-Site 913 (Figs. 6.5, 6.6) is 17 km offset from this line. The age information were extrapolated to this profile at CDP 1252 in a water depth of 3350 m.

All west-east profiles show prograding sequences at the East Greenland shelf. These units are clearly visible in the subsurface layers over a distance of about 60 km west of the shelf break on profile 20030390. P-wave velocities in the upper part of the shelf region determined by seismic velocity analysis range between 1.8 and 2.3 km/s. Velocities of 3.2 and 4.3 km/s were used from seismic refraction measurements (Voss and Jokat, 2007) for depths greater than 2 km in this region to convert the data to depth. According to the naming of Hamann et al. (2005) and Tsikalas et al. (2005) the line crosses the Thetis Basin (CDP 6250 to 9200) and the Danmarkshavn Ridge (Fig. 6.5; CDP 9200 to 10170). The top of basement is well imaged on the eastern side of the high. A basement structure at CDP 3090 has a northwest-southeast extent of approximately 32 km and an elevation of 750 m. It is found at a water depth of 2500 m. Furthermore, this structure shows a break at 3000 m (CDP 4170), and we observe a small irregularity in the basement topography (CDP 1870 - 2525) on its eastern flank. Hinz et al. (1987) called this step (Figs. 6.5, 6.6; CDP 4170) Greenland Escarpment (GE). A transparent reflection character is found within deeper sediment layers between CDP 1050 and the eastern end of the profile, defined as slump deposits by Myhre et al.(1995) on the basis of drilling results at ODP site 913. The position of the continent-ocean transition zone (COT; Fig. 6.1) was adopted from Voss and Jokat (2007) for profiles 20030390, 20030350, 20030550 and 20030560. Between profile 20030390 and profile 20030350 the position of the COT was interpolated, whereas profiles 20030380, 20030360 and 20030350 show weak internal basement reflections, which we interpret as SDRs (Fig. 6.5). The appearance of SDRs on these profiles confirm the eastern termination of the COT by Voss and Jokat (2007). On profile 20030370 SDRs could not been identified.

Sediments with progradation and aggradation sequences on profile 20030380 show great similarity to profile 20030390. Also the deeper transparent part of the Thetis Basin and Danmarkshavn Ridge looks identical with line 20030390 (Fig. 6.5). More eroded topsets are visible towards the shelf break in contrast to the northernmost profile 20030390 (Fig. 6.7). Profiles 20030390 and 20030380 show well-developed prograded sequences caused by advances and retreats of the East Greenland ice shield (Fig. 6.7). More aggradation is imaged on the outer shelf on profiles 20030370, 20030360 and 20030350 (Fig. 6.7). Diffuse layering with changing inclination within the upper shelf sediments are observed on profile 20030350 (Figs. 6.5, 6.7; CDP 7400 - 8650). Furthermore the top of acoustic basement was well imaged along these lines. The basement high north of profile 20030370 is located in the lower slope region (Figs. 6.3, 6.5). On profiles 20030370 and 20030360 the basement high is visible beneath the shelf break (Fig. 6.5, 6.6: 20030370: CDP 5000 - 6450; 20030360: CDP 3000 - 4250), and is overlain by around 2 km of sediment in comparison to the northern profiles with up to 500 m of sediment.

The two southernmost profiles (Figs. 6.1, 6.8) located in the prolongation of Godthåb Gulf (20030560) and the Kejser Franz Joseph Fjord (20030550) are identical with seismic refrac-

5.4 Profile description

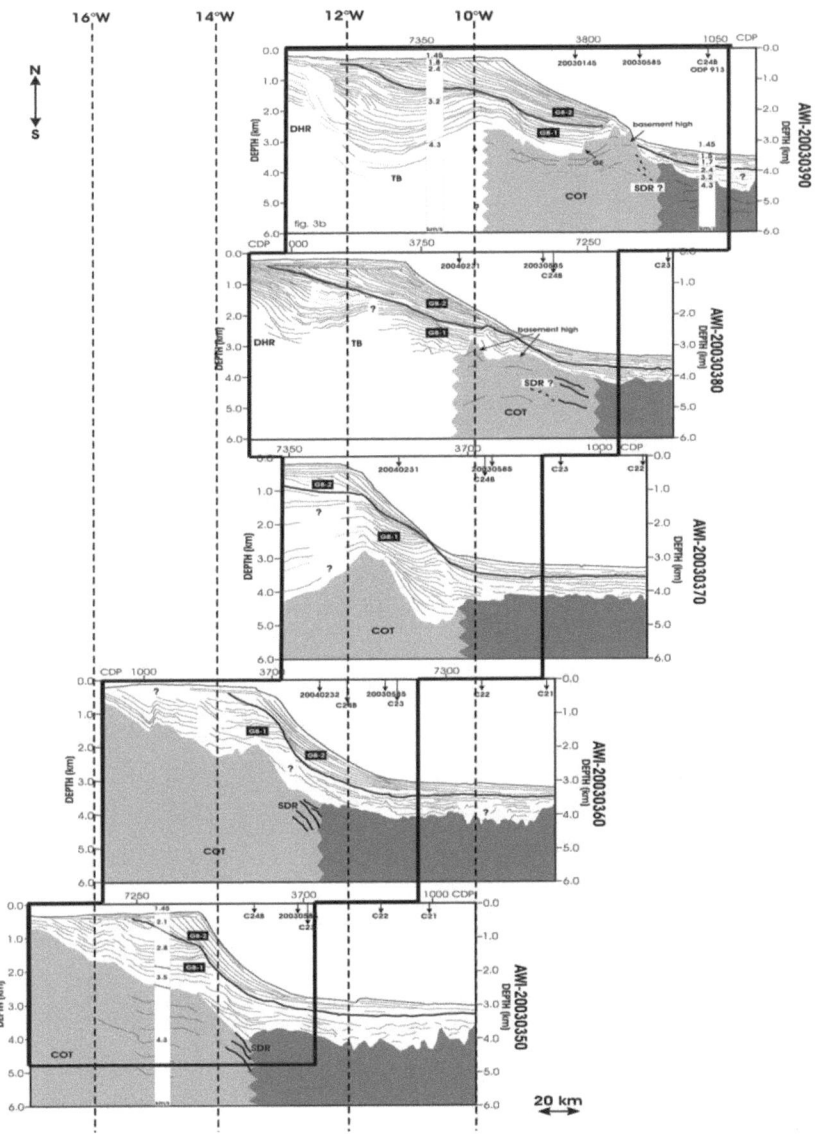

Figure 5.5: Line drawings of profiles AWI-20030390, AWI-20030380, AWI-20030370, AWI-20030360 and AWI-20030350 (southern Greenland Basin) arranged from north to south across the East Greenland continental margin (longitudes 10°W up to 16°W). The numbers at the black arrows represent the ties with crossing profiles at this position. The location of the magnetic spreading anomalies is marked. C24B = 56 Myrs., C23 = 51 Myrs., C22 = 49 Myrs., C21 = 47 Myrs (Gradstein et al., 2004). ODP-Site 913 is located in the deep sea area on profile AWI-20030390. The heavy black box represents parts of profiles shown in figure 6.6. The basement is divided into oceanic basement (dark gray) and continent-ocean transition zone (COT, light gray). DHR = Danmarkshavn Ridge, TB = Thetis Basin, SDR = seaward dipping reflectors, GE = Greenland Escarpment, GB-1 and GB-2 = seismic units.

51

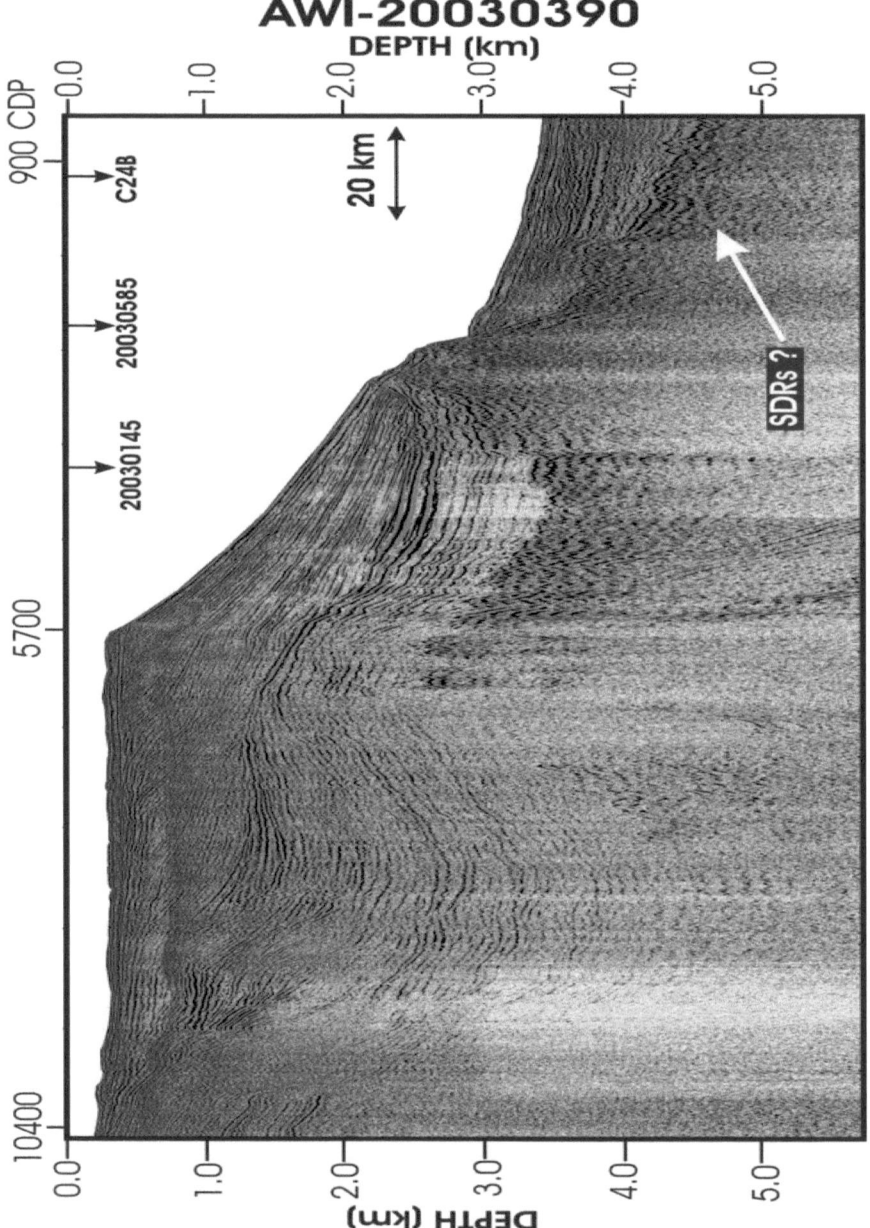

Figure 5.6: Seismic depth section of profile AWI-20030390 recorded with the 3000 m long streamer, illustrating the outer shelf, slope and deep sea of the Greenland Basin. The dotted white line represents the interpreted horizon, which divides the sediment package into GB-1 and GB-2, SDRs = seaward dipping reflectors.

5.4 Profile description

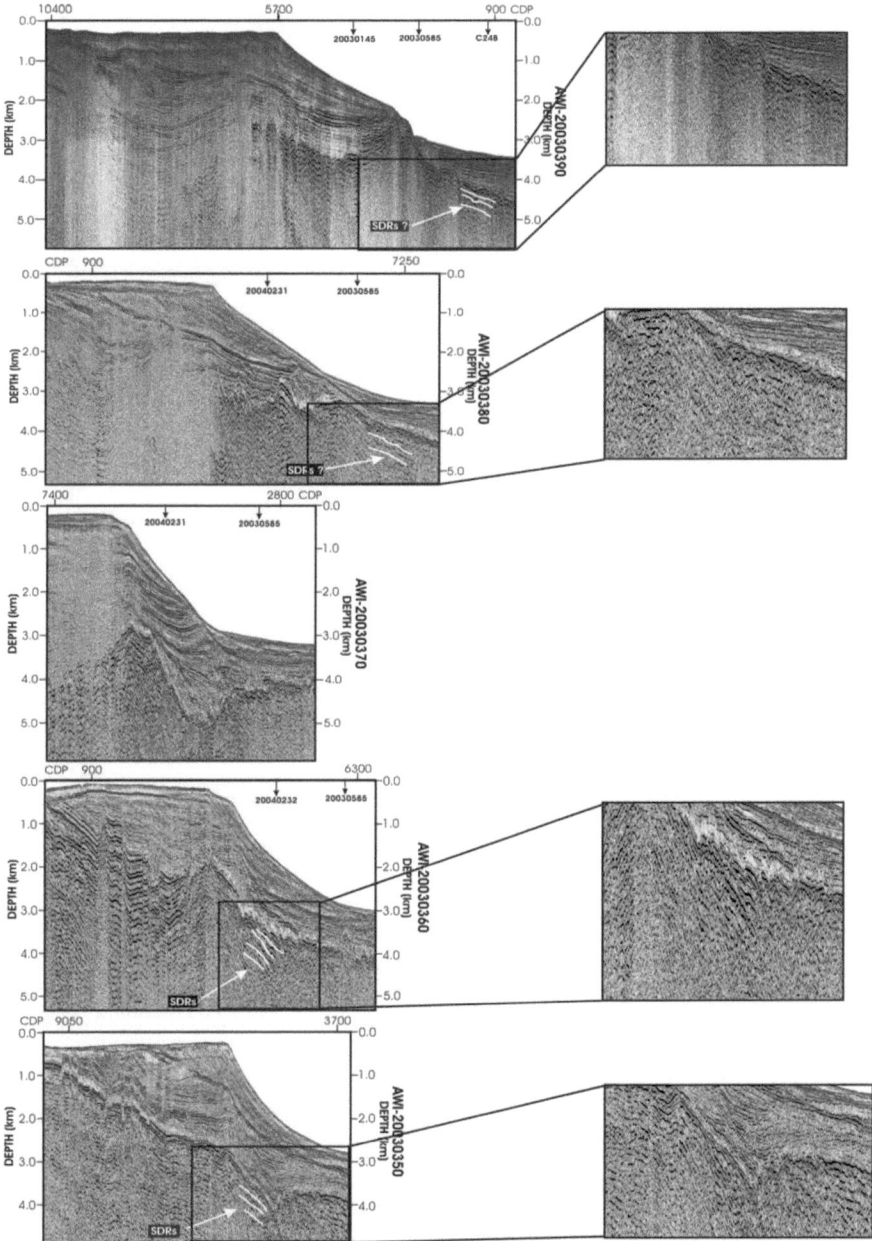

Figure 5.7: Parts of profiles AWI-20030390, AWI-20030380, AWI-20030370, AWI-20030360 and AWI-20030350. Part of the section is zoomed in to show the area of the SDRs in greater detail.

A seismic study along the Northeast Greenland margin

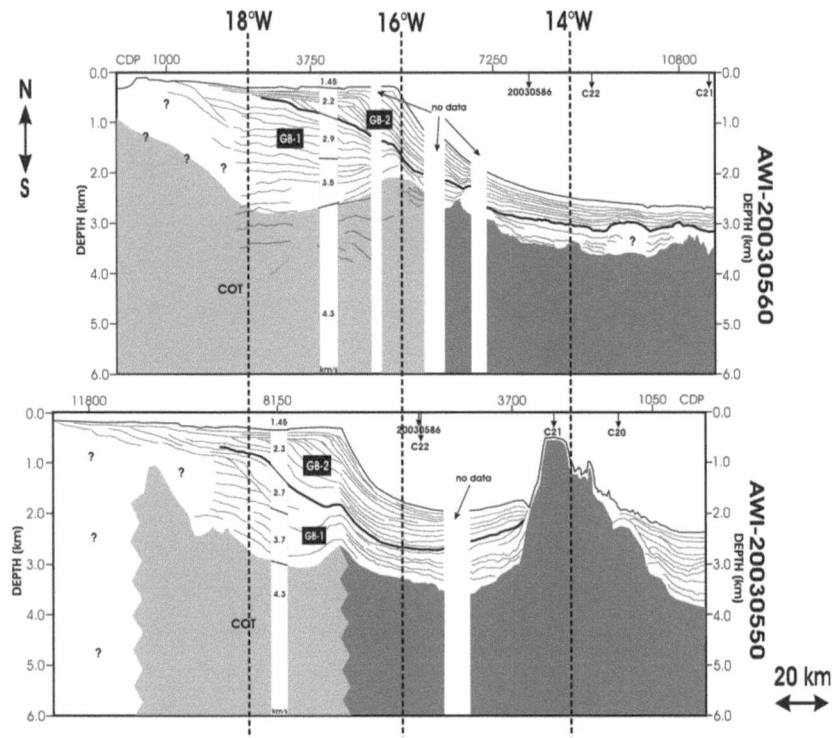

Figure 5.8: Line drawings of profiles AWI-20030560 and AWI-20030550 (southern Greenland Basin) arranged from north to south and orientated on longitudes 14°W up to 18°W. The numbers at the black arrows represent the crossing profiles at this position. The location of the magnetic spreading anomalies is marked. C22 = 49 Myrs., C21 = 47 Myrs., C20 = 43 Myrs. (Gradstein et al., 2004). The basement is divided into oceanic basement (dark gray) and continent-ocean transition zone (COT, light gray).

tion profiles described by Voss and Jokat (2007). Not only the location of the top of basement within the shelf were adopted from the deep sounding data, but also sediment velocities (2.7 to 3.7 km/s) for units deeper than 1 km to convert the seismic data into depth. In contrast to profile 20030560 a prominent basement structure in the oceanic domain is imaged on profile 20030550 (Fig. 6.8; CDP 1580 - 3350). This ridge and the outer shelf forms a 68 km wide basin (CDP 3350 up to 6370) filled with 1600 m of sediment. The north-south trending profiles 20030585 and 20030586 (Figs. 6.9, 6.10) as well as profiles 20030145, 20040230, 20040231 and 20040232 (Figs. 6.11, 5.12) are important tie lines for the stratigraphic correlations within the network. The top of oceanic basement (Fig. 6.9) could not be detected with great certainty at the southern termination of profile 20030586. Great variations in sediment thickness exist, and range between 300 m at the northern end of profile 20030585, up to approximately 2000 m at CDP 9340 on profile 20030586 (Fig. 6.9). Identifiable reflected signals from sonobuoy data along profiles 20040230 (Fig. 6.11; SB 19) and 20040231 (Fig. 6.11; SB 20) were additionally used to control the correlation of horizons, and check the seismic velocities for time to depth conversion on crossing profiles. The sediments from lower

5.4 Profile description

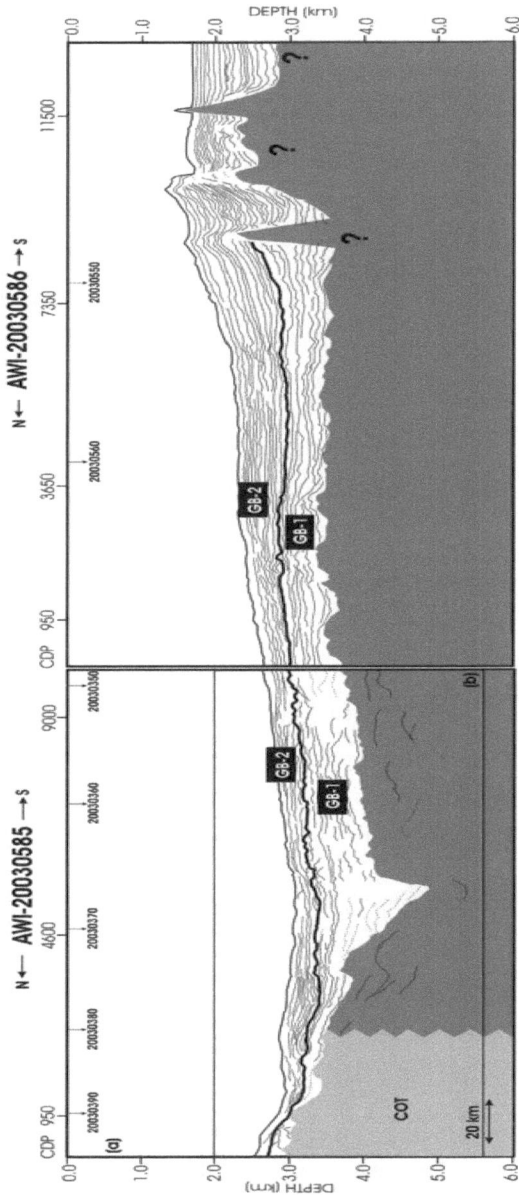

Figure 5.9: Line drawings of profiles AWI-20030585 and AWI-20030586, north-south orientated profiles along the East Greenland margin (Fig. 6.1). The location of the crossing profiles are marked. Dark gray show oceanic basement, GB-1 (0 - 3 Ma) and GB-2 (3 - 56 Ma) = seismic units, COT = continent-ocean transition zone.

A seismic study along the Northeast Greenland margin

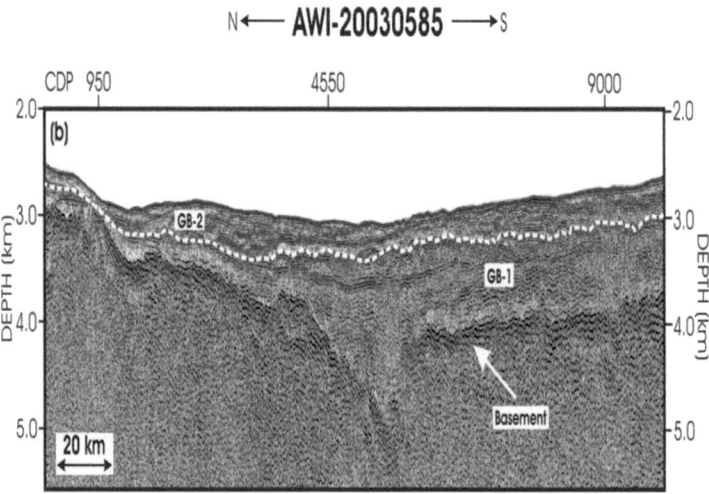

Figure 5.10: Seismic depth section of profile AWI-20030585 recorded with the 600 m long streamer. The white dotted line represents the interpreted horizon, which divides the sediment package into GB-1 and GB-2.

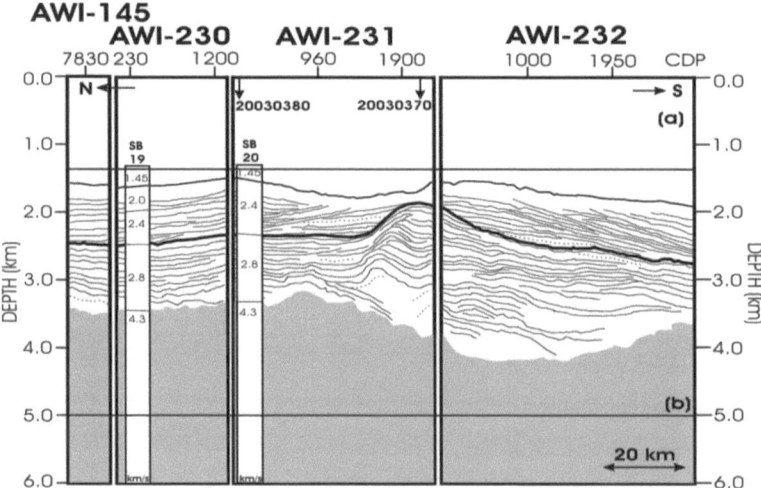

Figure 5.11: Line drawings of profiles AWI-20030145, AWI-20040230, AWI-20040231 and AWI-20040232 situated parallel to the East Greenland slope in north-south direction (Fig. 6.1). Sonobuoy 19 (SB 19) and sonobuoy 20 (SB 20) show seismic velocities in km/s used for the depth conversion. The basement (light gray) represents the continent-ocean transition zone. GB-1 and GB-2 = seismic units.

5.5 Stratigraphy

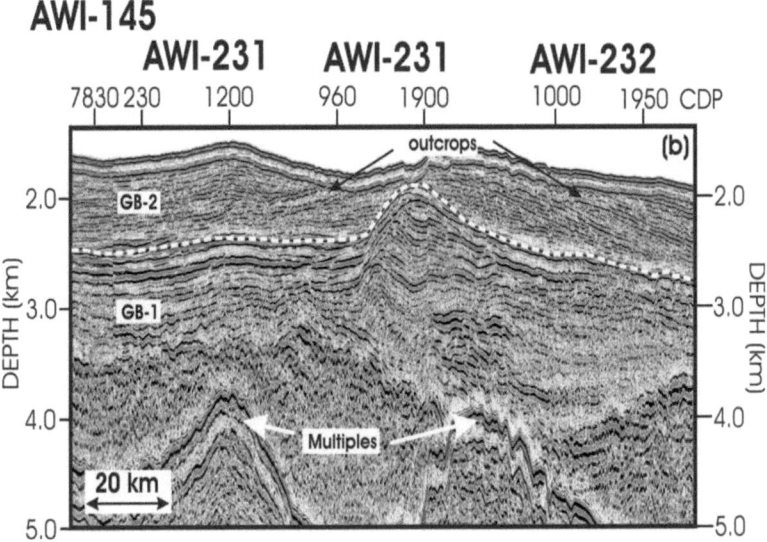

Figure 5.12: Seismic depth section of profiles AWI-20030145, AWI-20040230, AWI-20040231 and AWI-20040232 recorded with the 600 m long streamer, illustrating the upper slope region. The white dotted line represents the interpreted horizon, which divides the sediment package into GB-1 and GB-2.

levels left and right of CDP 1950 (Fig. 5.12; 20040232) crop out or almost reach the seafloor (Fig. 5.12; 20040231; CDP 875 to 1345 and 20040232; CDP 400 to 2520). The layering of the sediments below 2.5 km is less disturbed.

5.5 Stratigraphy

5.5.1 Available age information

In 1993, the RV "Joides Resolution" voyaged the East Greenland margin. During this cruise (Leg 151), seven deep sites were drilled (Myhre et al., 1995). Site 913 was drilled in the deep Greenland Basin south of the Greenland Fracture Zone at 75°29N, 6°96W (Fig. 6.1). The hole was drilled in a water depth of 3318 m, and penetrated to 770 m below the seafloor (m.b.s.f.) (Figs. 6.5, 6.8). The oldest magnetic seafloor spreading anomaly in this region is Anomaly 24B = ~56 Myrs (Talwani and Eldholm, 1977).
The uppermost Unit I (Fig. 5.13; Unit IA and IB) includes large dropstones ranging from 4 to 28 per meter. This interval from 0 - 144 m.b.s.f. is dominated by glaciomarine materials (Myhre et al., 1995) with an age of Pliocene to Quaternary (0 - 3 Myrs.). Their modell results in a sedimentation rate of 4.8 cm/kyr (Unit I). The description of the upper and lower boundary of Unit II is problematic, because of the poor recovery (2.8 %). The thickness of Unit II is given with 235 m (144 - 379 m.b.s.f.). Below the top of the boundary Unit II only a few large dropstones occur, and the lower boundary is placed at the lowest occurence of beds of clayey, silty, and sandy muds (379 m.b.s.f.). The few large dropstones in the interval between 220 - 423 m.b.s.f. are originally interpreted to be drilling contaminations (Myhre et

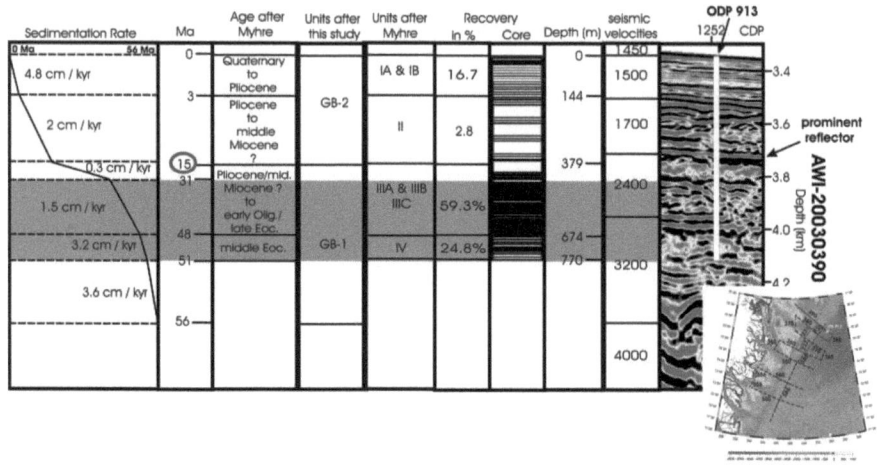

Figure 5.13: Stratigraphic model for the deep-sea part of the Greenland Basin. Information on ODP drill hole 913 was taken from Myhre et al. (1995) and Eldrett et al. (2004). The age classification by Eldrett et al. (2004) is imaged as gray area. Seismic velocities are determined on the basis of seismic velocity analysis. Definition of unit GB-1 and unit GB-2 was carried out on one prominent horizon visible on profiles located in the southern Greenland Basin. The emphasized age of 15 Ma represents the age of the prominent reflector in the drill hole depth of 360 m.b.s.f. The inset map shows the profiles, which were used to correlate the age information of site 913 onto the shelf.

al., 1995). Following this interpretation, the boundary between Unit II and Unit III represents a change from primarily nonglacial to glacial sediment deposition. The authors also describe the lack of dropstones below the upper boundary of Unit II as little ice-rafting activity during the middle Miocene to Pliocene. The middle Miocene age were dated at sediments containing an assemblage of moderately well-preserved Miocene diatoms and radiolarians (Myhre et al., 1995). Due to this information it is difficult to estimate an age of the Unit II/Unit III boundary (379 m.b.s.f.).

The deeper lithologies (Unit III and Unit IV) are described as massive and laminated silty clay and clay, with a maximum age of middle Eocene determined at 770 m.b.s.f. (Myhre et al., 1995). The top of oceanic basement was not reached. With a biostratigraphic analysis on dinoflagellate cysts (Eldrett et al., 2004) it was possible to setup a new biostratigraphy for the Eocene-Oligocene interval between 425 m.b.s.f. and 722 m.b.s.f. The good recovery (Fig. 5.13, 84.1 %) and the detailed age analysis within this interval represents the most complete and best-preserved record of the Eocene and Oligocene in the Northern Hemisphere high latitudes. Thus, the age of Unit IIIB, IIIC and a part of Unit IV (674 - 722 m.b.s.f.) have been specified with 35.3 Ma to 39.5 Ma (Unit IIIA), 39.5 Ma to 48 Ma (Unit IIIB) and 48 - 51 Ma (Unit IV, Eldrett et al., 2004).

5.5.2 Age correlation and seismic stratigraphy

Previous age information on the East Greenland shelf (Tsikalas et al., 2005) are based on reflection character, regional considerations and onshore Greenland geology (Stemmerik, 1993). Our correlation is mainly based on marker horizons and reflection character using

ODP site 913. At CDP 1252 on profile 20030390 some seismic horizons are observable. The units after Myhre et al. (1995) can be identified at this location. A direct correlation of these horizons into the shelf region is prevented by a basement high in the slope region, since the sediments vanish almost above this structure. Therefore, we acquired tie lines to transfer the age of sediment from site 913 onto the shelf. The correlation starts in the deep sea of profile 2000390 and was carried out along lines 20030385, 20030380, 20030585, 20030360, 20040232, 20040231, 20040230 and 20030145 (Fig. 5.13: inset). East of the high the sediment structure looks chaotic and is interpreted as slump area, following the assumption of Myhre et al. (1995). A prominent horizon at a depth of 360 m.b.s.f. has very high amplitudes, and could be identified outside the slump area along lines 20030585 and 20030586. The other lithostratigraphic units of site 913 were difficult or impossible to correlate into our network. They only have continuity close to the drill site. Away from the site these signals vanish or become highly subdued, which makes a sound stratigraphic correlation impossible. The prominent horizon is marked with a thick black line in figures 6.5, 6.8, 6.9 and 6.11 and by a white dashed line in figures 6.6, 6.10 and 5.12.

Age information combined with seismic reflection data at the borehole location are summarized in figure 5.13. The gray area in figure 5.13 marks the part of the borehole with good recovery and precise age information. For shallower parts a lot of data gaps exist.

For the depth conversion of the seismic section we have used only the seismic velocities from the velocity analysis of the seismic reflection profile. Due to the results of the correlation we have merged units IA, IB, II (Myhre et al. 1995) to unit GB-2 and units IIIA, IIIB, IIIC, IV (Myhre et al., 1995) to unit GB-1. According to the found Miocene diatoms and radiolarians at a depth of 375 m.b.s.f. (Myhre et al., 1995), an age of 15 Myrs could be determined for the GB-2/GB-1 boundary (Fig. 5.13).

The depth differences of the lower boundary between GB-2/GB-1 (375 m.b.s.f.) and the depth of the prominent reflector (360 m.b.s.f.) may be due to inaccuracies in final depth conversion, and the extrapolation of age information from site 913 to line 20030390 (17 km). Because of the dropstones at larger depths (308 and ~400 m.b.s.f., Myhre et al., 1995), and the classification of the boundary of nonglacial to glacial sediments (unit II/unit IIIA) by Myhre et al. (1995), we suggest that sediment at 360 m.b.s.f. were transported in middle Miocene times under glacially conditions.

The result of our correlation shows a division of sediments in the Greenland Basin and the adjacent East Greenland shelf in primarily nonglacial sediments deposited between 15 and 56 Ma, and glacial sediments accumulated between 15 Ma and present times. Based on this classification we follow the speculation of Winkler et al. (2002), that ice-rafting has probably occured since the middle Miocene.

5.6 Interpretation

5.6.1 Basement structures

A basement high was discovered beneath the East Greenland margin at the same position (75°43N, 7°58W) by Hinz et al. (1987). Eldholm and Windisch (1974) have introduced a high overlain by sediments of 0.6 s thickness (75.3°N, 10.5°W and 76.4°N, 5°W). A bit south seismic refraction data from Mutter and Zehnder (1988) show also a basement high, which is landward terminated by an escarpment. No geological age for this feature is available. It was unknown if this structure represents one large slope parallel feature or consists of a series of local highs. From the new seismic network and the additional bathymetric information, we note that the highs do not occur at a constant water depth. The structural appearance

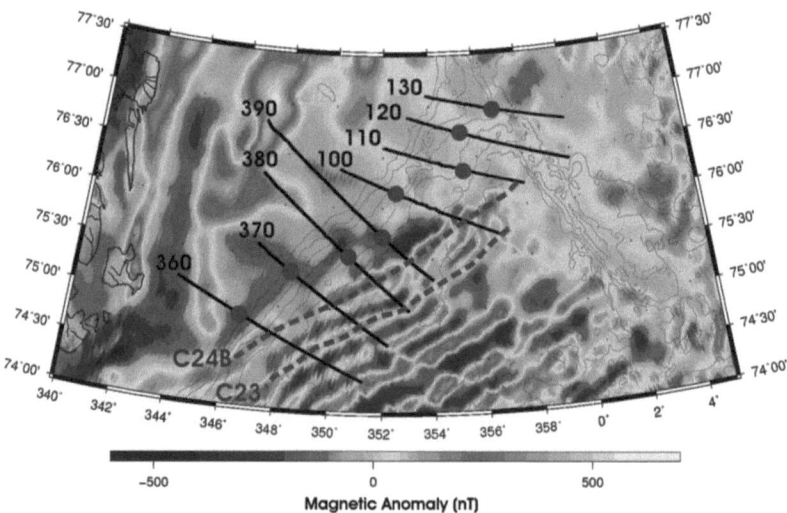

Figure 5.14: Magnetic anomalies off East Greenland. The black lines show the location of the seismic profiles across the East Greenland continental margin. Dashed gray lines are the oldest spreading anomalies found in the Greenland Basin (C 24B = 56 Myr, C 23 = 51 Myr, Escher et al., 1995; Eldholm et al., 2002a). Red dots locate the basement highs. GFZ = Greenland Fracture Zone.

also varies from line to line. Figures 6.3 and 6.5 summarizes the results of the E-W seismic profiles. The line drawings show that the basement high is present between lines 20030130 and 20030360. However, the structure is not aligned along a straight line. The bathymetric data do not provide any evidence for fault-controlled offsets that might separate it into several fragments.

Figure 5.14 shows the relationship of seafloor spreading anomalies between 74°54N and 77°N to the seismic profiles of this study. The oldest anomaly identified in this region is anomaly C24B (Talwani and Eldholm, 1977). In figure 5.14, all crossings of the basement structure are located landward of the oldest magnetic anomaly. The nature and age of the crust is unknown between the oldest anomalies and the basement high. However, there are several interesting observations:

- The distance between the basement high and chron C24B increases towards the north. This might suggest that this structure formed diachronously.

- As mentioned earlier, the continous structure is not a straight lineament and occurs today at different water depths.

- An increasing dip of sediments on both sides at the basement high is not visible, which could point to an uplift after the break-up.

- Some profiles (Fig. 6.7: 20030390, 20030380, 20030360, 20030350) show evidences for SDRs seaward of the prominent basement structure.

5.6 Interpretation

- The seismic data show no layering within the basement, which argues against a sedimentary origin (Tsikalas et al., 2005). Furthermore, seismic velocities of the rocks at the top of the high is around 4.3 km/s (Voss et al., submitted 2007)

Mutter and Zehnder (1988) calculated a velocity of 4.0 km/s for the upper part of the high at 75°40N and 7°50W. Similar velocities are found by Voss et al. (submitted). Voss et al., (submitted) interpret the basement high on the basis of seismic refraction data, as basaltic material.
Tsikalas et al. (2002) suggest a segmentation of the East Greenland margin along fracture zones, which extend from oceanic crust onto the shelf. E.g., the proposed existence of the Bivrost Fracture Zone at 74°40N, 12°W. The seismic network especially in the deep sea provides no evidence for the existence of any major fracture zones. Thus, our data do not support the proposed margin segmentation. These findings are similar to results of Olesen et al. (2007), who identified navigation errors in the old magnetic data, which were interpreted as fracture zone. An new survey confirmed that the magnetic field is not offset by fracture zones.
From several publications we know that Tertiary magmatism accompanied the final stage of rifting, and the opening of the North Atlantic in Early Eocene times (White and McKenzie, 1989; Skogseid et al., 2000; Eldholm and Grue, 1994). The different rifting histories north and south of the Kong Oscar Fjord led to the assumption, that pre-existing crustal structures had a controlling character for magmatic depositions (Schlindwein and Jokat, 1999). Volcanic material intruded into the sedimentary basins and large amounts of flood basalts were extruded in the southern region around Scoresby Sund (Escher and Pulvertaft, 1995). Hints for increased magmatic melt production in the lower crust in this region are given by Schlindwein & Jokat (1999), Voss and Jokat (2007). Deep seismic data in the prolongation of the Kong Oscar Fjord, Kejser Franz Joseph Fjord and Godthåb Gulf is interpreted as magmatic underplating of the East Greenland continental crust. This body vanishes in the northwards direction, and to north of Shannon Island, only slightly intruded lower crust is visible (Voss et al., submitted). The deep seismic data indicate that the volcanic activity decreases towards the Greenland Fracture Zone. If such a basement high existed in the south it might be buried by younger volcanic material, which erupted prior or during the breakup. Because of the lower amount of erupted material in the northern part of the Greenland Basin, this high might be visible. Another explanation is that the style of volcanism changed in the north and produced material only within a small corridor of the COT. SDRs on the seaward flank of the high postulated by Hinz et al. (1987) are similar to dipping reflectors at the conjugate Vøring Plateau. SDR units on the Norwegian side were drilled and tholeiitic material were recorded (Eldholm et al., 1987). We assume the same composition for the conjugate East Greenland SDRs. Based on these observations we follow the original interpretation of Hinz et al. (1987) and Voss & Jokat (2007) and consider the high as a volcanic structure. This volcanic structure might been emplaced during the initial extensional phases of the opening of the Greenland Basin. The question, if it is a continuous structure or consist the basement high of several fragments, is difficult to answer, as a denser seismic reflection network is needed. However, we favour the interpretation that the basement highs form a prominent, diachronuously continuous structure with a N-S extent of approximately 360 km in different water depth, based on the structural symmetry of N-S and similar seismic refraction velocities (Voss et al., submitted; Mutter & Zehnder 1988). Another prominent basement structure is the GFZ, as imaged on profile 20030120 (Fig. 6.3). The GFZ separates the Greenland Basin from the Boreas Basin and has developed during the formation of the Greenland Basin since ~ 56 Ma (Talwani and Eldholm, 1977). The GFZ represents a

plate tectonic feature initiated at the transition between rifted and sheared continental margin segments (Faleide et al., 1993). During the Eocene, the proto Greenland-Senja Fracture Zone (pGSFZ) comprised the Senja Fracture Zone along the SW Barents Sea margin and the Greenland Fracture Zone along the southwest foot of the elongate, monolithic Greenland Ridge (Tsikalas et al., 2002). Faleide et al. (2001) speculate that the Greenland Ridge does not resemble a typical oceanic fracture zone and suggested that it may contain a continental sliver. Representative seismic velocities of the crust are not available, and until now it is not clear, if the GFZ is made up of oceanic crust, continental crust or a mixture of both.

A plate tectonic model for the evolution of the Greenland (GFZ)-Senja (SFZ) Fracture zone - Greenland Ridge system is introduced by Tsikalas et al. (2002). In this modell NW-SE movements dominate. The 360 km long volcanic structure found north of 75°N along the East Greenland margin is NE-SW orientated, and seems to have an independent formation of the GFZ. Hence, the basement high on profile 20030130 (Fig. 6.3, CDP 3060 - 3560) may mark the location, where the GFZ becomes decoupled from the East Greenland margin. In the absence of any deep seismic sounding data it can only be speculated if this process was accompanied by excessive volcanism or not.

5.6.2 Total sediment thickness

The overall trend is an increase of total sediment thickness (Fig. 5.15) from the deep sea towards the East Greenland shelf break. The maximum sediment thickness of 3032 m is present on the continental slope at 75°18N.

All W-E profiles longer than 100 km in the northern Greenland Basin show a marked decrease in sediment thickness, to less than 250 m, across the basement highs in the slope region. These prominent basement structures are thus, likely to have formed an obstacle to sediment transport from the shelf into the abyssal plain. In general, the sediment thickness is approximately 1000 m in the Greenland Basin, and on average 1800 m in the southern Boreas Basin (Fig. 5.15). The seismic data show approximately twice as much sediment in the Boreas Basin than in the Greenland Basin. One reason for this could be the absence of a barrier like the basement high on the continental slope. Alternatively, or additionaly the GFZ may have acted as a barrier to southward directed current transport of suspended material originating from the northern margins of Greenland (77°N and 80°N). In support of this view, it is known that persistent currents are flowing southwards parallel to the margin, in this case the East Greenland current (Rudels et al., 2002). A part of this current separates at the GFZ and enters the Boreas Basin (Rudels et al., 2002). We assume during the separation the current velocity decreases and more material could be accumulated north of the GFZ. Also the much wider East Greenland shelf north of the GFZ (Fig. 6.1), could explain that the erosion provided more material than south of 76°N. South of 77°N well developed prograding sequences on the outer shelf and high velocities (1.8 - 2.3 km/s) near the top of the seafloor indicate that the shelf was glacially eroded. So we guess the major source areas for sediments are the Greenland mainland and the shelf. Obviously, the basement highs and the GFZ play an important role in the distribution of the sediment in the investigated area.

5.6.3 Glacial - Preglacial sediments

Preglacial sediment (GB-1) thickness (Fig. 5.16) was calculated as the depth difference of the top of acoustic basement and the top of unit GB-1 (15 to 56 Ma). Younger sediments (up to 15 Ma) are represented by the difference between the top of unit GB-1 and the seafloor

5.6 Interpretation

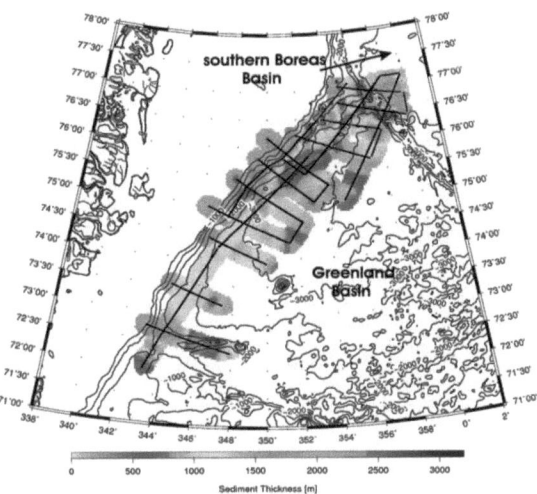

Figure 5.15: The sediment thickness grid shows a general trend of sediment distribution close to the Greenland margin. Only the parts of the seismic reflection network is shown (black lines), which provide information for this compilation. Bathymetric contours are plotted with a spacing of 1000 m.

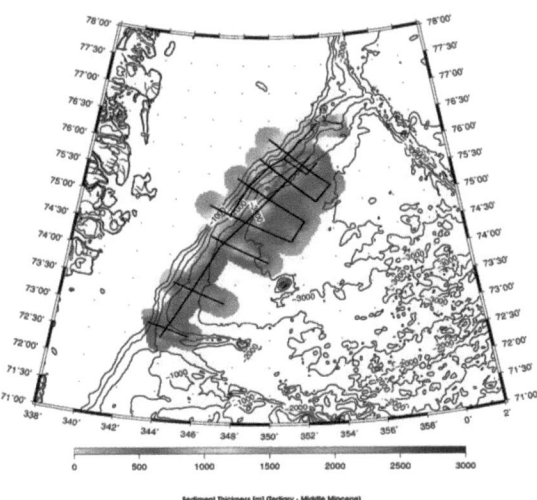

Figure 5.16: Gridded version of Middle Miocene - present sediment distribution (GB-1) in the Greenland Basin. Only the parts of the seismic reflection network is shown (black lines), which provide information for this compilation. Bathymetric contours are plotted with a spacing of 1000 m.

63

A seismic study along the Northeast Greenland margin

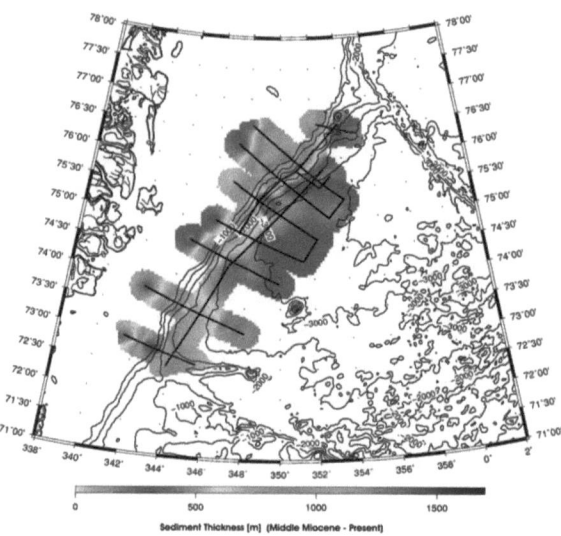

Figure 5.17: Gridded version of Middle Miocene - Tertiary sediment distribution (GB-2) in the Greenland Basin. Only the parts of the seismic reflection network is shown (black lines), which provide information for this compilation. Bathymetric contours are plotted with a spacing of 1000 m.

(Fig. 5.17; GB-2).
The thickness of GB-1 (Fig. 5.17) varies between 34 m at the top of the basement high on profile 20030380 and 2720 m at the slope on profile 20030370 (Fig. 6.5). Most of the accumulated sediments were found along the East Greenland slope between 74°30N and 76°30N. The basement highs visible on profiles 20030390 and 20030380 (Fig. 6.5) represent a barrier for the ice rafted material. The top of the high shows almost no sediment deposits before 15 Myrs, and the thickness of unit GB-1 decreases from approximately 1000 m in the west of the highs to 400 m in the east of the highs. Also noticeable is the continuously transition (72°30N - 75°N) of the sediment deposits from the slope to the deep sea where no basement highs could be observed. Sediments within layer GB-1 in the deep Greenland Basin between 73°45N and 75°30N increase from 200 m in the north to 1000 m in the south caused by the disappearance of the basement highs in southern direction. The thickness of glacial sediment deposits is mapped in figure 5.17, and range between 110 m and 1670 m. The sedimentation rates (GB-2) are in general similar in the deep Greenland Basin (2.8 cm/kyrs). However, south of 73°N the sedimentation rates increase to 4.8 cm/kyrs. The transported sediments in the prolongation of the Kejser Franz Joseph Fjord (20030550) is representative for the tremendous sediment transport by glaciers. The largest thicknesses of glacial sediments of 1600 m are found beneath the shelf break. Here, prograding sequences developed in the late Cenozoic as a consequence of advances and retreats if glaciers/ice streams on the outer shelf (Figs. 6.5, 6.6).

A comparison between sediments, which are younger (GB-2) and older (GB-1) than 15 Myrs show extremly low sedimentation rates for the interval between the continental breakup and the beginning of progradation (GB-1). An average sedimentation rate for the deep sea was calculated with 1.2 cm/kyrs, only 40 percent of glacial sediment deposits. That means, either most of the older sediments were transported away by deep sea currents or the glaciation of the northern hemisphere has resulted in massive increase of erosion of the Greenland mainland and shelf starting 15 Ma. By means of the broad prograding sequences on the East Greenland shelf over a distance of more than 60 km we favour the last interpretation. In our interpretation the sequences are caused as a consequence of sea level changes and glacial erosion. The extend of such an early glaciation is speculative. However, we propose that already in Middle Miocene times some localised glaciers existed.

5.7 Conclusion

The new seismic profiles in the northern part of the Greenland Basin indicate that a basement high exists south of the Greenland Fracture Zone. This basement structure is clearly visible between 74°54N to 77°N and shows a N-S extent of approximately 360 km. By considering magnetic spreading anomalies and the position of seaward dipping reflectors in the southern Greenland Basin, the structure is located within the continent-ocean transition zone, and seems to have developed parallel with the rifting in this area. We propose that the structure is a volcanic feature with a slightly older age than 56 Ma. A continuation of this basement structure north of the GFZ seems not to be very probable, but the database in the Boreas Basin does not allow us to rule this out.

Furthermore, our data support the location of the COT termination more or less underneath the present day shelf edge as proposed independently by deep seismic profiles. The location of this basement structure, together with weak volcanic seaward dipping sequences support the location of the continent-ocean transition termination as determined by other deep seismic sounding studies. The multichannel seismic lines acquired in 2003 were used to compile a set of maps that provide insight into the sedimentary evolution of this basin. Problems in correlating the deep basin stratigraphy onto the shelf was caused by the presence of a 360 km long volcanic basement structure. A thickness of up to 1 km of glacial deposited sediments were found beneath the shelf break. Prograding sequences are observed over a distance up to 80 km. They started to develop since Middle Miocene times. Sediment thickness values of 1000 m are found in the Greenland Basin, compared to 1800 m in the adjacent Boreas Basin. This higher sediment accumulation might be explained by the separation of the East Greenland current north of the GFZ, the wider shelf area along the Boreas Basin and the absence of the volcanic structure in the slope region. We assume that sediment deposition in the Greenland Basin is mainly influenced by mass transport from the continental slope into the deep basin by fast travelling ice-streams and gravity-driven processes. Evidences for current controlled deposition are obvious. Based on these results we assume that the glaciation of East Greenland caused a massive erosion of the mainland, and might have been started already some 15 Ma.

Finally, a denser seismic network around site 913 is needed in order to correlate more units along the margin for more detailed description of the sedimentary history.

5.8 Acknowledgements

We are grateful for the excellent support of the captain and crew of the RV "Polarstern". This research was partly funded by StatoilHydro and the Deutsche Forschungsgesellschaft. We thank R. Mjelde, University of Bergen, Norway for providing the streamer system for this cruise.

5.9 References

Dowdeswell, J.A. (1996), *Large-scale sedimentation on the glacier-influenced Polar North Atlantic Margins: long-range side-scan sonar evidence*, Geophysical Research Letters 23, 3535-3538.

Eldholm, O., Windisch, C.C. (1974), *Sediment Distribution in the Norwegian-Greenland Sea*, Geological Society of America Bulletin, Vol. 85, 1661-1676.

Eldholm, O., Thiede, J., Taylor, E., et al. (1987), *Norwegian Sea*, Proceedings of the Ocean Drilling Program, Initial Reports Vol. 104.

Eldholm, O. and Grue, K., (1994), *North Atlantic volcanic margins: Dimensions and production rates*, Journal of Geophysical Research 99(B2): 2955 - 2968.

Eldrett, J.S., Harding, I.C., Firth, J.V., Roberts A.P. (2004), *Magnetostratigraphic calibration of Eocene-Oligocene dinoflagellate cyst biostratigraphy from the Norwegian-Greenland Sea*, Marine Geology, 204, 91-127.

Eldrett, J.S., Harding, I.C., Wilson, P.A., Butler, E., Roberts, A.P. (2007), *Continental ice in Greenland during the Eocene and Oligocene*, Nature, Vol. 446.

Escher, J. and Pulvertaft, T. (1995), *Geological Map of Greenland, 1:2500000*, Geological Survey of Greenland.

Ewing, J.I. and Ewing, W.M. (1959), *Seismic refraction measurements in the Atlantic Ocean Basins, in the Mediterranean Sea, on the Mid-Atlantic Ridge, and in the Norwegian Sea*, Geol. Soc. America Bull. 70, p. 291-318.

Faleide, J.I., Vågnes, E. and Gudlaugsson, S.T. (1993), *Late Mesozoic-Cenozoic evolution of the south-western Barents Sea in a regional rift-shear tectonic setting*, Mar. Petrol. Geol. 10, 186-214.

Faleide, J.I., Tsikalas, F. and Eldholm, O. (2001), *Regional riftshear tectonic setting and Late Cretaceous-early Tertiary events linking the Lofoten-Vesterålen and SW Barents Sea margins (NE Atlantic)*, in S. Roth and A. Rüggeberg (eds), 2001 MARGINS Meeting, Schriftenreihe der Deutschen Geologischen Gesellschaft.

Gradstein, F., Ogg, J. and Smith, A. (eds.) (2004), *A geological time scale 2004*, Cambridge University Press, Cambridge, United Kingdom (GBR).

Hamann, N.E., Whittaker, R.C. and Stemmerik, L. (2005), *Geological development of the Northeast Greenland Shelf*. In: Doré, A.G. & Vining (2005), B.A. Petroleum Geology: North-West Europe and Global Perspectives-Proceedings of the 6th Petroleum Geology conference, Volume 2, 887-902, Geological Society, London.

5.9 References

Helland, P.E., Holms, M.A. (1997), *Surface textural analysis of quartz sand grains from ODP Site 918 off the southeast coast of Greenland suggests glaciation of southern Greenland at 11 Ma.* Palaeogeogr. Palaeoclimatol. Palaeoecol. 135, 109-121.

Hinz, K., Mutter, J.C., Zehnder, C.M. & Group, N.S. (1987), *Symmetric conjugation of continent-ocean boundary structures along the Norwegian and East Greenland margins*, Mar. Petrol. Geol. 3, 166-187.

Jokat, W., Berger, D., Bohlmann, H., Helm, V., Hensch, M., Jousselin, D., Klein, C., Lensch, N., Liersch, P., Martens, H., Medow, A., Micksch, U., Rabenstein, L., Salat, C., Schmidt-Aursch, M. & Schwenk, A. (2004), *Marine Geophysics. Reports on Polar and Marine Research 475*, Bremerhaven, 11-34.

Jokat, W., Behr, Y., Birnstiel, H., Gebauer, A., Göling, K., Günther, D., Martens, H., Lensch, N., Raabe, W., Schmidt-Aursch, M., Schroeder, M. Spengler, T. (2005), *Marine Geophysics*, Reports on Polar and Marine Research 517, Bremerhaven, 27-40.

Larsen, H.C., Saunders, A.D., Clift, P.D., Beget, J., Wei, W., Spezzaferri, S. (1994), *ODP Leg 152 Scientific Party, 1994. Seven Million Years of Glaciation in Greenland*, Science 264, 952-955.

Moore, G.T., Hayashida, D.N., Ross, C.A. and Jacobsen, S.R., (1992a), *Paleoclimate of the Kimmeridgian/Tithonian (Late Jurassic) world: I. Results using a general circulation model.* Paleogeography, Paleoclimatology, Paleoecology, 93: 113-150.

Moore, G.T., Sloan, L.C., Hayashida, D.N. and Umrigar, N.P., (1992b), *Paleoclimate of the Kimmeridgian/Tithonian (Late Jurassic) world: II. Sensivity tests comparing three different paleotopographic settings*, Paleogeography, Paleoclimatology, Paleoecolgy, 95: 229-252.

Mutter, J. C. & C. M. Zehnder, (1988), *Deep crustal and magmatic processes: The inception of seafloor spreading in the Norwegian-Greenland Sea*, Geol. Soc. Spec. Publ. London 39, 35-48.

Myhre, A.M. and Thiede, J., Firth J.V. et al. (1995), *North Atlantic-Arctic Gateways, Proceedings of the Ocean Drilling Program*, Initial Reports Vol. 151.

Ó Cofaigh, C., Dowdeswell, J.A., Grobe, H. (2001), *Holocene glacimarine sedimentation, inner Scoresby Sund, East Greenland: the influence of fast-flowing ice sheet outlet glaciers*, Marine Geology, 175, 103-129.

Olesen, O., Gellein, J., Håbekke, H., Kihle, O., Skilbrei, J.R. & Smethurst, M.A. (1997), *Magnetic anomaly map, Norway and adjacent ocean areas*, Scale 1:3 million, Geological Survey of Norway.

Olesen, O., Ebbing, J., Lundin, E., Mauring, E., Skilbrei, J.R., Torsvik, T.H., Hansen, E.K., Henningsen, T., Midbøe, P., Sand, M. (2007), *An improved tectonic model for the*

Eocene opening of the Norwegian-Greenland Sea: Use of modern magnetic data, Marine and Petroleum Geology, Vol. 24, 55-66.

Ostenso, N.A., and Wold, R.J. (1971), *Aeromagnetic survey of the Arctic Ocean: Techniques and interpretations*, Marine Geophysical Research, v1, p. 178-219.

Planke, S. and Alvestad, E. (1999), *Seismic Volcanostratigraphy of the extrusive breakup complexes in the northeast Atlantic: Implications from ODP/DSDP drilling*, Proc. ODP, Sci. Results, 163.

Rudels, E., Fahrbach, E., Meincke, J., Budéus, G., Eriksson, P., (2002), *The East Greenland Current and its contribution to the Denmark Strait overflow*. Journal of Marine Science, 59: 1133-1154.

Schlindwein, V. and Jokat, W. (1999), *Structure and evolution of the continental crust of northern east Greenland from integrated geophysical studies*. Journal of Geophysical Research 104(B7): 15227 - 15245.

Skogseid, J., Planke, S., Faleide, J.I., Pedersen, T., Eldholm, O. & Neverdal, F., (2000), *NE Atlantic Continental rifting and volcanic margin formation*. In: Nottredt, A. (ed.) Dynamics of the Norwegian Margin. Geological Society, London, Special Publications, 167, 295-326.

Stemmerik, L. (1993), *Depositional history and petroleum geology of Carboniferous to Cretaceous sediments in the northern part of East Greenland*. In: Vorren, T.O. et al., (Eds.), Arctic Geology and Petroleum Potential. Norwegian Petroleum Society (NPF), Special Publication 2. Elsevier, Amsterdam, pp. 67-87.

Talwani, M. and Eldholm, O. (1977), *Evolution of the Norwegian-Greenland Sea*, Geological Society of America Bulletin 88, 969-999.

Tsikalas, P., Eldholm, O., Faleide, J.I., (2002), *Early Eocene sea floor spreading and continent-ocean boundary between Jan Mayen and Senja fracture zones in the Norwegian-Greenland Sea*, Marine Geophysical Research, 247-270.

Tsikalas, F., Faleide, J.I., Eldholm, O. and Wilson, J. (2005), *Late Mesozoic-Cenozoic structural and Stratigraphic correlations between the conjugate mid-Norway and NE Greenland continental Margins, Geological development of the Northeast Greenland Shelf*. In: Doré, A.G. & Vining, B.A. (2005) Petroleum Geology: North-West Europe and Global Perspectives-Proceedings of the 6th Petroleum Geology conference, Volume 2, 785-801, Geological Society, London.

Vanneste, K., Uenzelmann-Neben, G., Miller H. (1995), *Seismic evidence of long-term history of glaciation on central East Greenland shelf south of Scoresby Sund*. Geomarine Letters 15, 63-70.

5.9 References

Voss, M. and Jokat, W. (2007), *Continent-ocean transition and voluminous magmatic underplating derived from P-wave velocity modelling of the East Greenland continental margin*. Geophysical Journal International, Volume 170, Issue 2, Page 580 - 604.

Voss, M., Schmidt-Aursch, M.C., Jokat, W. (submitted 2007), *Variations in magmatic processes along the East Greenland volcanic margin*. Geophysical Journal International.

White, R.S. and McKenzie, D. (1989), *Magmatism at rift zones: The generation of volcanic continental margins and flood basalts*, Journal of Geophysical Research 94(B6): 7685 - 7729.

Winkler, A., Wolf-Welling, T.C.W., Stattegger, K., Thiede, J. (2002), *Clay mineral sedimentation in high northern latitude deep-sea basins since the Middle Miocene (ODP Leg 151, NAAG)*, Int. J. Earth Sci. 91, 133-148.

Ziegler, P.A. (1990), *Geological Atlas of Western and Central Europe*. Geological Society Publishing House, London, U.K., 239 pp.

6 Paper 2

Sediment deposition in the northern basins of the North Atlantic and characteristic variations in the shelf sedimentation along the East Greenland margin

Daniela Berger and Wilfried Jokat

Marine and Petroleum Geology (2008), Vol. 26, Issue8, pp. 1321-1337.

accepted 2009 April, received 2009 March 3, in original form 2008 October 14

6.1 Abstract

New seismic data off East Greenland were acquired in the summer 2002, between 77°N and 81°N, north of the Greenland Fracture zone. The data were combined with results from the Greenland Basin and ODP site 909, and indicated a pronounced middle Miocene unconformity, within the deep sea basins between 72°N and 81°N. Seismic unit NA-1 was dated to consist of sediments older than middle Miocene and unit NA-2 contains of sediments younger than Middle Miocene. A further classification of the fine bedded sediment succession in the Molloy Basin resulted in a subdivision of four units. A comparison of volume estimations and sediment thickness maps between 72°N and 81°N show differences in the sediment accumulation in the Greenland, Boreas and Molloy basins. Important controls on the variation of sediment accumulation were the different opening times of the basins, as well as tectonic conditions and varying sources of sediment transport. Due to prominent basement structures and the varying reflection character of the sediments along the entire East Greenland margin, we defined an age model of shelf sediments on the basis of similar internal geometries and known results from other regions. Therefore, the seismic sequences on the shelf up to an age of middle Miocene have been divided into three sub-units along the East Greenland margin: middle Miocene-middle late Miocene (SU3), middle late Miocene-Pleistocene (SU2), Pleistocene (SU1). The differences in the geometry of the sequences give reason to speculate about different sediment transport processes like ice-stream and ice-sheet related sedimentation along the East Greenland margin. Due to the Greenland Inland-ice borderlines, we assume the glaciers between the Scoresby Sund and 68°N did not reach the shelf break. Strong ice stream related sedimentation was observed on the shelf adjacent to the Greenland Basin.

Key words: glacial sediments; prograding foresets; sediment transport; seismic reflection

6.2 Introduction

The initial opening of the Norwegian Greenland Sea took place during early Eocene times (Talwani and Eldholm, 1977). The first sea-floor spreading anomalies in the Boreas Basin were dated back to 35 Ma and the opening of the Molloy Basin began 21 Ma (Ehlers and

Jokat, 2008). There is little known about the timing of the opening of the deep-water connection between the Arctic Ocean and the northern North Atlantic. This is important to know, because the deep-water formation and exchange could be an indicator for changes in global climates. Results from the Integrated Ocean Drilling Program (IODP) Expedition 302 (ACEX) in 2004 show a transition from poor oxygenated to fully oxygenated conditions occuring during the later part of the early Miocene, in the Arctic Ocean (Moran et al., 2006; Jakobsson et al., 2007). Moran et al. (2006) attribute this change in the ventilation regime to the opening of the Fram Strait. Ehlers and Jokat (2008) postulate for the initial opening of the Fram Strait, an age of 17-18 Ma (middle Miocene) based on the analysis of magnetic seafloor spreading anomalies. The important question whether there is a relationship between the opening of the Fram Strait and the onset of the glaciation of the Northern Hemisphere remains unclear.

The glacial history of the Northern Hemisphere is a subject of controversy. Accordingly, accurate reconstructions of Greenland climate history is an important precondition for our understanding of the sources of global climate change and sea level variations. Least understood are the speculations about the onset of the glaciation of East Greenland and vary from Plio/Pleistocene to Eocene (Winkler et al., 2002; Tsikalas et al., 2005; Eldrett et al., 2007). The absence of drill holes on the East Greenland shelf further complicate the understanding of the glacial history of this island.

Present geophysical studies were concentrated on the Southeast Greenland margin (Larsen et al., 1994; Planke and Alvestad, 1999) to cover the volcanic complexes and to retrieve information about the breakup scenario in this region. Seismic measurements were carried out during the seasons 1980-1982 by the GEUS (geological Survey of Denmark and Greenland). ODP Site 914 is located on the South East Greenland shelf above the flood basalt area (Planke and Alvestad, 1999). The tie between the borehole and seismic reflection profile GGUi/82-02, show an onset of prograding foresets on the Southeast Greenland margin in middle late Miocene (around 7 Ma, Larsen et al., 1994; Larsen et al., 1999) times. Tsikalas et al. (2005) has made the first seismostratigraphic correlation for the Northeast Greenland shelf, based on reflection character, regional considerations and onshore Greenland Geology (Haman et al., 2005; Stemmerik et al., 1993). The model from Tsikalas et al. (2005) shows an age of Plio/Pleistocene for the base of progradation at 76°N. However, Berger and Jokat (2008) have made the first correlation from the deep Greenland Basin towards the shelf on the basis of the ODP drill Site 913 (Myhre et al., 1995). Their results show an onset of the progradation in middle Miocene times and the unconformity at the base of progradation is interpreted as consequence of strong sea level changes and glacial erosion. Other speculations depend on analyses of drill hole material from deep sea positions. By means of ice rafted debris and proxy data of the ODP Site 909 and Site 908, Winkler et al. (2002) suggest a cooling phase during the middle Miocene. Results of analyses of ODP Site 913 in the deep Greenland Basin report stratigraphically extensive ice-rafted debris, including macroscopic dropstones, in late Eocene to early Oligocene sediments deposited between about 38 and 30 million years ago (Eldrett et al., 2007). The authors assume the sediment rafting is mainly caused by glacial sea ice and suggest East Greenland as the likely source. The latest IODP ACEX drilling results derive from a recent drilling expedition to the Lomonosov Ridge in the Arctic Ocean, show gneiss dropstones in an undisturbed sediment section, dated of around 45 Ma and interpreted as ice-rafted debris. At the same site a 26 Myr long hiatus (Backmann et al., 2005) separates the middle Eocene (44.4 Ma) sediments from early Miocene (18.2 Ma) sediments. Moran et al. (2006) define this interval as overlapping with the timing of a global shift from largely ice-free, warm world with high relative sea level to a world characterized by the climate-modulated waxing and waning of ice sheets. At the end of this interval in

the early Miocene times the abundance of dropstones and sand suggests that sea ice and icebergs, calved from glaciers, were present in the Arctic Ocean (Moran et al., 2006).
Results from the Norwegian margin, Southeast Greenland margin and Antarctic margin show that glacial shelf sediments are mostly represented by prograding sequences. Grounded ice transports a large volume of sediments across the shelf edge and unsorted sediments are deposited on the outer shelf and slope. The repeated advance of ice sheets to the shelf edge during glacial periods, results in a cyclicity in sediment supply. These advances are viewed as the mechanism for the development of the prograding sequences. The start of the progradation is often interpreted as the start of the regular advances of the grounded ice to the shelf edge (e.g. Cooper et al., 1991; Barker and Camerlenghi, 2002; Nielsen et al., 2005). Another reason for prograding clinoforms can be relative sea level changes (Vail et al., 1977).
A new seismic dataset between 77°N and 80°30′N (Fig. 6.1) covers the Molloy Basin and the Boreas Basin, as well as the adjacent slope area in the west and parts of the East Greenland shelf (Fig. 1). The data were acquired in the summer 2002 with "RV Polarstern" during the expedition ARK XVIII/2 by the Alfred Wegener Institute for Polar and Marine Research (Jokat et al., 2003). 5847 km of seismic reflection data were gathered to promote the understanding of the sedimentation history along the East Greenland margin specifically in the north. The interpretation of this seismic dataset combined with information from ODP Site 909, and compared with results from previous studies along the entire East Greenland margin, provided a new insight into the glacial and sedimentary history along the margin.

6.3 Data acquisition and processing

A seismic reflection survey between 77°N and 81°30′N was carried out with a cluster of 8 x 3l of airguns, in the summer 2002. Due to heavy ice conditions during this campaign a streamer with a length of 600 m was used. The streamer was provided with 96 active channels, and had a hydrophone group spacing of 6.25 m. The shot interval was 15 s, which is equivalent to a shot distance of 35-40 m. Seventeen sonobuoys were additionally deployed along the East Greenland margin to complement the network.
The multichannel seismic data were demultiplexed, edited, sorted and binned into 25 m spaced CDPs. Before stacking, the data were band pass filtered with 15-90 Hz. The short streamer provided only limited velocity information for depth conversion, because of the small offset compared to the water depth (>1500 m). Representative stacking velocities (± 50 m/s) for this region were derived from sonobuoy data. They were used to define a simple but regionally consistent model of three layers plus basement. The velocities range between 1800 m/s for the uppermost sediment layer, and up to 3300 m/s at the top of the basement. After processing, the calculated sediment thicknesses were gridded with the surface utility from Generic Mapping Tools (GMT). To analyse the volume of the sediments in the Molloy, Boreas and Greenland Basin the areas of sediment accumulation were divided into grid cells with a size of 0.5x0.5 degrees.

6.4 ODP Site 909

Several sites were drilled during ODP cruise Leg 151 in the summer 1993 with the RV "Joides Resolution" on the East Greenland margin. Site 908 and site 909 (Fig. 6.1) were drilled in the deep sea at 78°23′N, 1°22′E (908) and 78°35′N, 3°4′E (909) in the northern North Atlantic. Site 909 was planned as the deep-water location in the Fram Strait on a small abyssal terrace located immediately to the north of Hovgård Ridge (Myhre et al., 1995). Further southwest

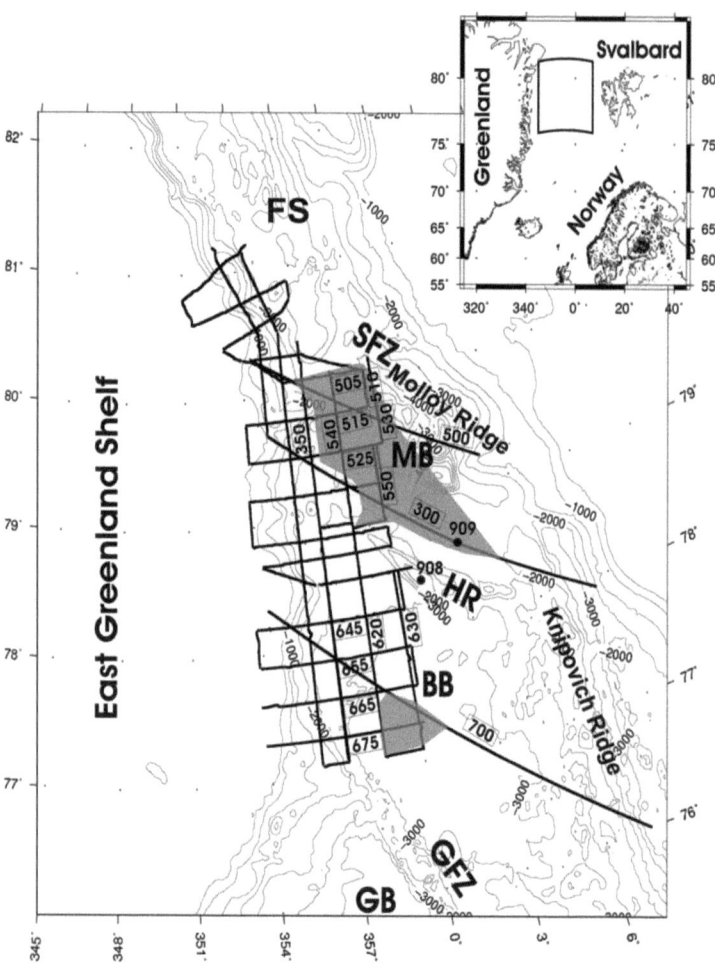

Figure 6.1: Map of the study area with bathymetric contours plotted with a spacing of 500 m. The black lines represent the locations of seismic reflection profiles in the Boreas and Molloy Basin. Labeled profiles will be described in subsection 4. The black dots represent the locations of ODP sites 909 and 908. SFZ, Senja Fracture zone; HR, Hovgård Ridge; GFZ, Greenland Fracture zone; MB, Molloy Basin; BB, Boreas Basin; GB, Greenland Basin. The gray area represents the location of thin-bedded sediments.

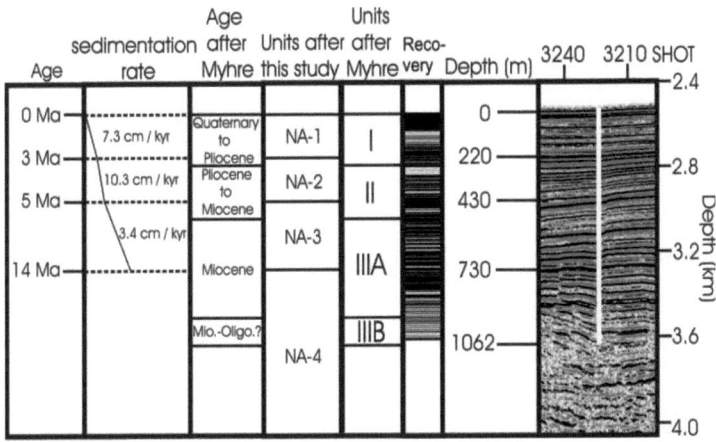

Figure 6.2: Stratigraphic model for the deep-sea part of the Boreas and Molloy basin. Information on ODP drill hole 909 was taken from Myhre et al. (1995). Seismic velocities are determined on the basis of seismic velocity analysis and sonobuoy data. Definition of unit NA-1 and unit NA-2 is based on a horizon of regional character. The left column (age) is only an approximation, based on the ages of Myhre et al. (1995). The emphasized age of middle Miocene represents the age of the prominent reflector in the drill hole depth of 730 m.b.s.f. provided that the sedimentation rate was nearly constant within interval IIIA.

site 908 was drilled on the top of the Hovgård Ridge, 50 km away from drill site 909. This core has a length of 1061 m and is situated in a water depth of 3590 m on profile 20020300 (Fig. 6.1). The age model is based on paleomagnetic data according to the time scale of Cande and Kent (1992). The general sediment composition is described as dark-gray silty clays (Myhre et al., 1995). The drilled sediment sections were divided into three main units (Fig. 6.2) and unit three was subdivided into A and B. Dropstones with a diameter > 1 cm were found from Pliocene to Quaternary in the interval of 0 - 248.8 m.b.s.f., which is named unit I (Fig. 6.2). The thickness of the lithological unit II is given with 269.5 m. (248.8 - 518.3 m.b.s.f.). The silty clay is more massive and interbedded with thinner layers of carbonate-rich clays. Dropstones were not found in this unit, which was dated from Pliocene to Miocene. The base of unit IIIA (518.3 - 923.4 m.b.s.f.) with a Miocene age is placed at the top of slump structures. The entire interval is of Miocene age. The silty and clayey lithology is characterized by meter-scale intervals of thin bioturbated layers and laminations (Myhre et al., 1995). The deepest unit IIIB (923.4 - 1061.8 m.b.s.f.) was dated as Miocene to upper upper Oligocene and has folded and deformed bedding (Myhre et al., 1995).

6.5 Results

The seismic profiles were subdivided into two parts, and will be described in this chapter. Whenever possible within these basins, a seismostratigraphic model was added. Eight profiles which cross the East Greenland shelf from west to east will be introduced, along with a classification of the shelf sediments.

6.5.1 Classification of seismic units

Lithological units by Myhre et al. (1995) have been correlated into the seismic network of the deep Molloy Basin. Within the deep Molloy Basin we find a thinly laminated sediment succession (Fig. 6.1: gray area) due to clear identifiable reflectors. The seismic resolution appears to be higher than in the adjacent regions (Figs. 6.2; 6.3; 6.4). A comparison of the seismic velocities from the thinly laminated sediment deposits with the transparent sediment succession is not possible because the short streamer (600 m) provided only limited velocity information. However, the area of thinly laminated sediments, however, is limited (Fig. 6.1), and a consistent age model with four units could not be created for the entire dataset between 77°N and 81°N. The seismic profiles in the Molloy Basin (Fig. 6.3) indicate that the area of fine layered sediment deposits is limited by several basement highs (Fig 6.3: 20020300, 20020505, 20020525). The seismic reflection character changes from fine layered reflection pattern (Fig. 6.3: 20020300: CDP 5000 to 10000) to sediment successions with transparent reflection character. One reflector is very dominant within the study area. This reflector is located within lithologic unit IIIA in a borehole depth of 730 m.b.s.f. and can be correlated through the entire data set. Myhre et al. (1995) suggests a Miocene age for this interval. Since the prominent reflector is situated in the middle of this interval (730 m.b.s.f.), we have estimated an age of middle Miocene for this reflector, based on the assumption of a constant sedimentation rate during this period. Therefore, the whole sediment package has been divided into two megasequences. This classification could be made for all the sediments in the Molloy and Boreas basin (77°N - 81°N). The sediments above the middle Miocene reflector are represented by seismic unit NA-2 (NA = North Atlantic) and the sediments between the middle Miocene and the top of the basement are represented by seismic unit NA-1. Unit NA-1 has an age of middle Miocene up to the beginning of the opening of the representative basins and unit NA-2 has an age range of middle Miocene to present day. The detailed seismostratigraphy of the four units (unit I to unit IIIB) was added where it was possible to subdivide the structure of the thinly laminated sediments (Figs. 6.3, 6.4, 6.5).

6.5.2 Molloy Basin

The Molloy Basin (Fig. 6.1) is bound in the south by the Hovgård Ridge (HR) and in the north by the Spitsbergen Fracture Zone (SFZ). Bathymetric data provide an average water depth of 3500 m for the central Molloy Basin.
Profile 20020300 (Fig. 6.3) has a total length of approximately 350 km, situated north of the Hovgård Ridge and orientated northwest-southeast. ODP Site 909 (Figs. 6.2, 6.3) is located at CDP 5675 in a water depth of 2520 m. The trench of the Knipovich Ridge at the eastern end of the profile, is also observable on profile 20020500, which is located around 100 km north of profile 20020300. Between CDP 1800 - 9750 (Fig. 6.3: AWI-20020300) a fine layered seismic reflection pattern of nearly 900 m is visible. Figure 6.1 images the area (gray), where thin-bedded sediments were found.
The seismic units were interpolated along crossing profiles of profile 20030300. The base of unit NA-2 on profile 20020300 (Fig. 6.3) is the boundary between the fine bedded reflection pattern and the completely transparent acoustically section below. At CDP 10000 (Fig. 6.3: 20020300) the basement crops out at the sea-floor and forms a boundary of the fine layered reflection pattern. The profile 100 km north (20020500) shows north-west of the Knipovich Ridge a fine layered reflection pattern between CDP 6450 and 4000 of around 1 km of sediment.
In the deeper parts below 3500 m depth on profile 20020300 and 4200 m depth on profile

6.5 Results

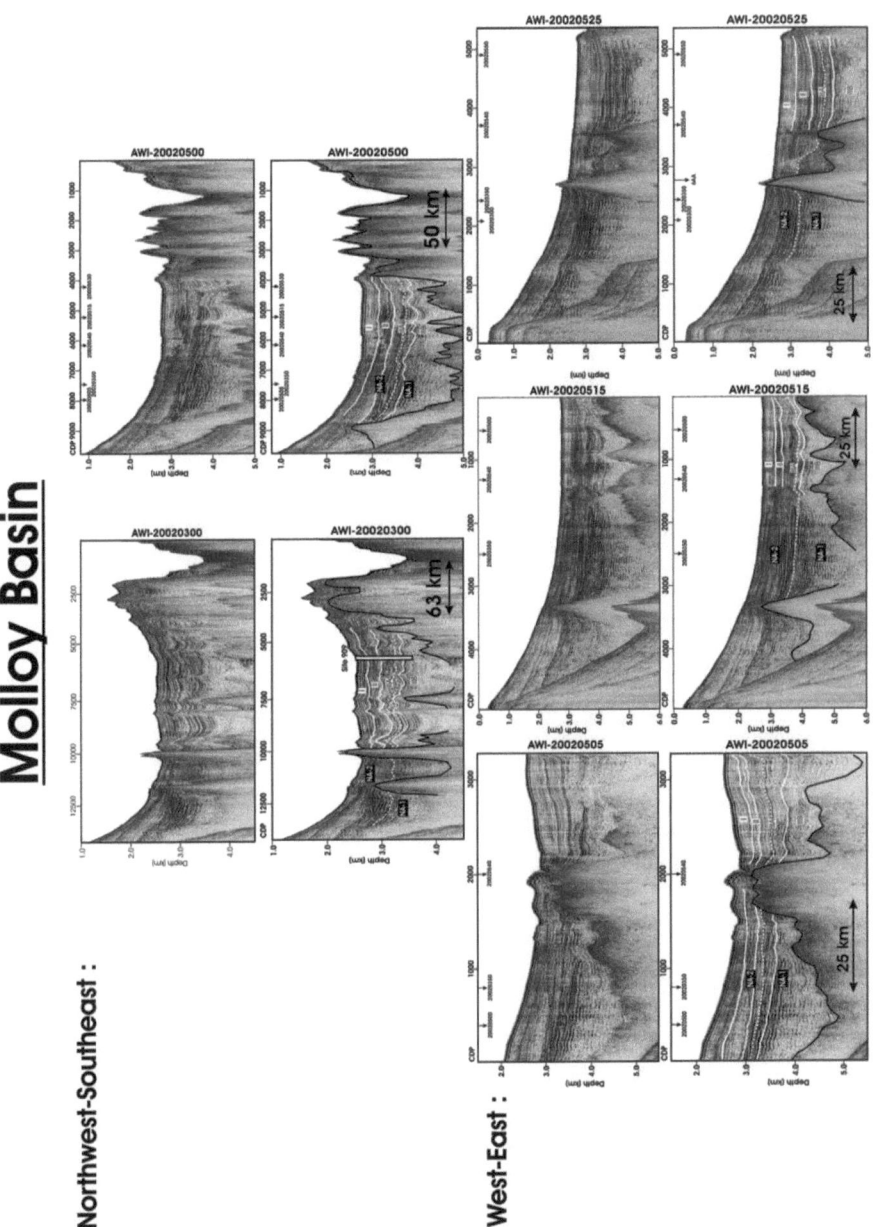

Figure 6.3: Seismic reflection profiles in the Molloy Basin. The location of the profiles is shown in figure 6.1. Numbers at the black arrows show the ties of the crossing profiles. 6AA on profile 20020525 represents the magnetic anomaly C6AA (21 Myr). NA-1 and NA-2 are seismic units, which are older and younger than middle Miocene respectively (white dashed lines demonstrate the transotion). The black line represents the top of the acoustic basement.

77

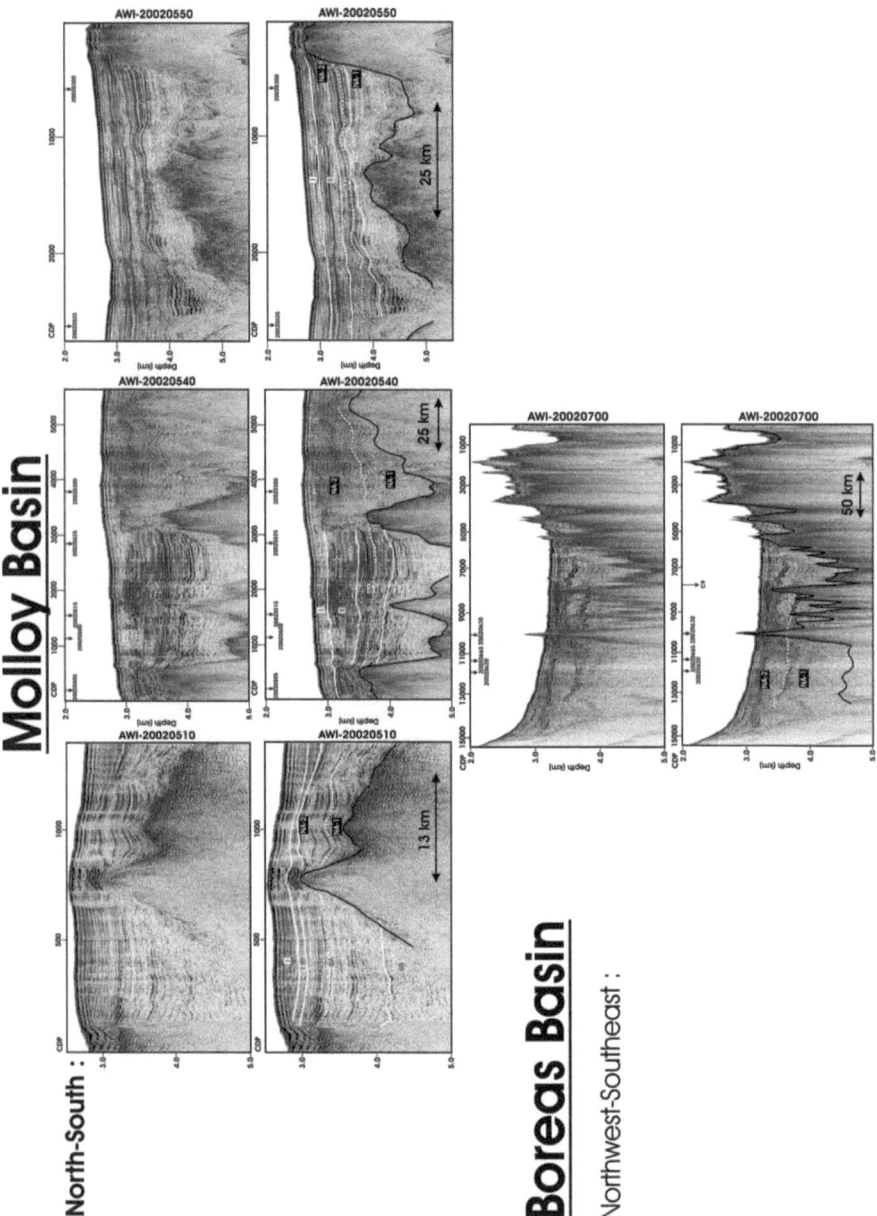

Figure 6.4: Seismic reflection profiles in the Molloy and Boreas Basin. The location of the profiles is shown in figure 6.1. The numbers with the black arrows show the ties of the crossing profiles. C9 on profile 20020700 represents a magnetic anomaly (28 Myr). NA-1 and NA-2 are seismic units, which are older and younger than middle Miocene (white dashed lines demonstrate the transition). The black line represents the top of the acoustic basement.

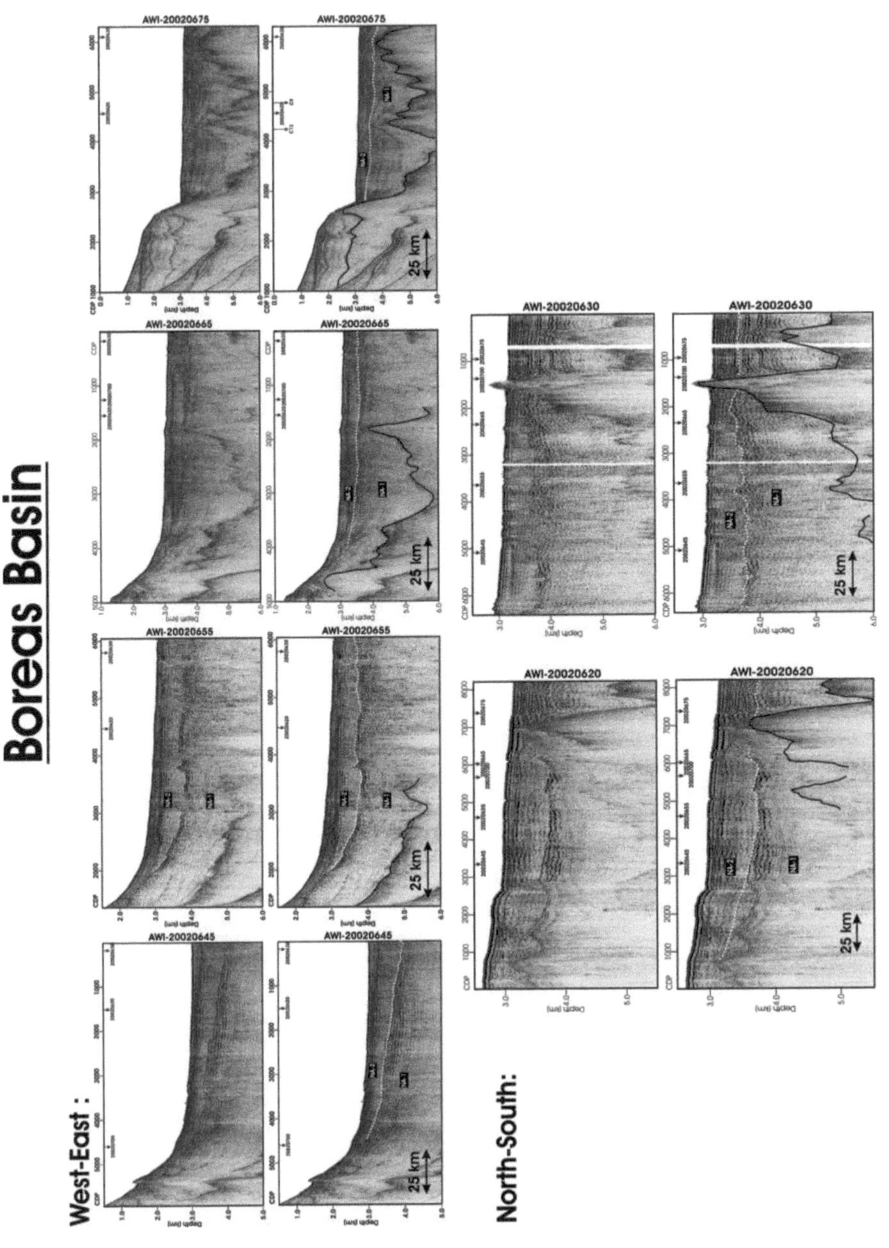

6.5 Results

Figure 6.5: Seismic reflection profiles in the Boreas Basin. The location of the profiles is shown in figure 6.1. The numbers with the black arrows show the ties of the crossing profiles. C9 and C13 on profile 20020675 represent magnetic anomalies (28 Myr and 33 Myr). NA-1 and NA-2 are seismic units, which are older and younger than middle Miocene respectively(white dashed lines demonstrate the transition). The black line represents the top of the acoustic basement.

20020500 a more transparent reflection character is observed up to the top of the acoustic basement. The lower sediments in unit NA-2 are characterized by high amplitudes compared to the deposits in the upper part of unit NA-1. On both northwest-southeast trending profiles the sediment deposition in the area between the upper slope region and the foot of the slope appears chaotic. This change from acoustically fine-layered sediments to chaotic deposits complicates an age correlation in the upper slope region.
The profiles 20020505, 20020515 and 20020525 are situated in west-east direction within the Molloy Basin (Fig. 6.3). These profiles show a rough basement structure in the lower slope region at a depth between 2 and 3 km. The top of this structure crops out at the sea-floor on profile 20020525, and almost reaches the sea-floor on profiles 20020505 and 20020515. Westwards of this structural high, 1 km of the upper part of NA-2, the sediments are characterized by a transparent reflection pattern. Below this section a sediment package with higher amplitudes is observed down to 4 km depth. Underneath, the continuous sediment structure represents an almost transparent reflection character. In the slope region, no reflections from the basement are visible because of strong sea-floor multiples (Fig. 6.3). On profile 20020515 high amplitudes can be found east of the high between CDP 2030 and 3175. The subdivision of the acoustically thinly laminated section into unit I to unit IIIB could be made at the eastern end of this profile (Fig. 6.3; CDP 20 to 1500) similar to profile 20020525 (Fig. 6.3; CDP 3700 to 5300). The north-south orientated profiles 20020510, 20020540 and 20020550 (Fig. 6.4) show fine layered sediments at the northern end of these profiles. The area of the fine layered sediments is marked with light gray in figure 6.1. Correlation with the magnetic isochrons dated by Ehlers and Jokat (2008), shows that the location of the oldest magnetic isochron 6AA (21 Myr) in the Molloy Basin (Gradstein et al., 2004) is coexistent with the western boundary of the gray highlighted area of the fine layered sediments (Fig. 6.1).

6.5.3 Boreas Basin

The Greenland Fracture zone (GFZ) separates the Greenland Basin from the Boreas Basin and the Boreas Basin is bound in the north by the Hovgård Ridge (Fig. 6.1). The Boreas Basin also shows a rough basement topography similar to the Molloy Basin. The identification of a continuous basement event was not possible in the deep sea part, along the three northern west-east profiles 20020645, 20020655 and 20020665 (Fig. 6.5), as well as between CDP 10550 and 15550 on the northwest-southeast profile 20020700 (Fig. 6.4), which cover the Knipovich Ridge (Fig. 6.1) in the southeast. The sedimentary strata in the Boreas Basin is divided into two the two sediment megasequences i.e. units NA-1 and NA-2. The high amplitude reflector is observed in the deep sea part at a depth between 3.5 and 4 km on all profiles (Fig. 6.5). Above this reflector, within unit NA-2, just beneath the seabed, a 300 m thick package of continuous reflections exist. The reflection pattern below the first 300 m.b.s.f. has been characterised as transparent. At depths of 4 km, underneath the prominent reflector (unit NA-1) the data also show a transparent reflection character up to the top of the basement, where it could be identified (Fig. 6.5; 20020645, 20020655, 20020665, 20020675). Also the north-south orientated profiles 20020620 and 20020630 (Fig. 6.5) support the characteristic description of the reflection pattern in the deep sea area of the Boreas Basin. However, at the southern end of these profiles - south of the structural basement high, visible in CDP 7100 and CDP 1350 (Fig. 6.5) - the structure changes from transparent into thinly laminated sediments with decreasing penetration depth. The southernmost west-east profile in the Boreas Basin (Fig. 6.5: 20020675) also shows a fine layered reflection pattern and a clearly identifiable basement structure in the deep sea part. In general, we could observe in the Boreas Basin a fine layered sediment structure between 77°N and 77°30'N (Fig. 6.1) close to the GFZ. The western boundary of this area is located between magnetic isochron 9 (28 Myr) and isochron 13 (33 Myr) in the southern Boreas Basin (Gradstein et al., 2004).

6.5.4 Prograding sequences on the East Greenland shelf

Eight seismic profiles from the Molloy Basin in the north to the Irminger Basin in the south (south of 65°N), provide an overview of the shelf progradation along the East Greenland margin. All profiles cover the shelf (Fig. 6.6) from west to east (profiles 1-3 only the outer shelf area) and show different seismic packages, which has been interpreted to present prograding clinoforms. The profiles south of the Scoresby Sund were made available by StatoilHydro (Fig. 6.6: KANU92E-5, KANU95-07) and the Geological Survey of Denmark and Greenland (Fig. 6.6: GGUi/82-02, GGUi/81-04). These seismic reflection profiles are available in two way travel time. For a better comparison of the shelf structures - from north to south - we have imaged also the northern profiles (Fig. 6.6: AWI-97250, AWI-97270, AWI-20030390, AWI-20030550) as time sections. The deposits have been divided into section with different characteristic geometries.

Classification of shelf sediments along the East Greenland margin

Because of the very rough basement in the north, fracture zones (GFZ, JMFZ) along the margin and changing sedimentation character, it is not possible to carry out a consistent seismostratigraphic correlation for the shelf sediments from north to south.
On the basis of the acoustic character and internal geometry of sediments, we have classified the accumulated shelf sediments along the entire East Greenland shelf. The outcome of this is a subdivision of the shelf sediments into three seismic units (SU). SU-2 and SU-3 represent sediment packages interpreted as prograding clinoforms. SU-1 is predominately characterized by exclusive horizontal reflectors interpreted as aggrading strata. Reflector R3 forms the base of SU-3 and separates prograding strata, above from non-prograding strata below. Reflector R2 separates SU-2 and SU-3 and provides the lower boundary of unit SU-2. The upper boundary of unit SU-3 is represented by reflector R1.

- Prograding units (SU-2 and SU-3):

  Based on the analysis of the acoustic character and internal geometry of seismic prograding sequences, we distinguish between prograding units: SU-2 and SU-3. On all profiles from north to south along the East Greenland shelf, the dip-angle of the foresets increases from the older SU-3 to the younger SU-2. In general, the geometry of unit SU-3 can be described as a mixture of truncated horizons and continuous sequences. Especially on profiles AWI-20030390 and KANU95-07 we can recognize some offlap terminations below an erosional surface (Fig. 6.6). Unit SU-3 was not identified on the two southernmost profiles GGUi/82-02 and GGUi/81-04 and on the northern profiles AWI-97270 and AWI-97250 (Fig. 6.6). The lower part of unit SU-3 seen on the KANU-MAS data (Fig. 6.6: KANU92E-5, KANU95-07), south of the Scoresby Sund show a very gently dip on the outer shelf. Around 15 km west of the shelf break, these sequences change into a more aggradational character. The greatest amount of sediment in this unit is visible on the East Greenland shelf adjacent to the Greenland Basin (Fig. 6.6: AWI-20030390, AWI-20030550).

Shelf sedimentation along the East Greenland margin

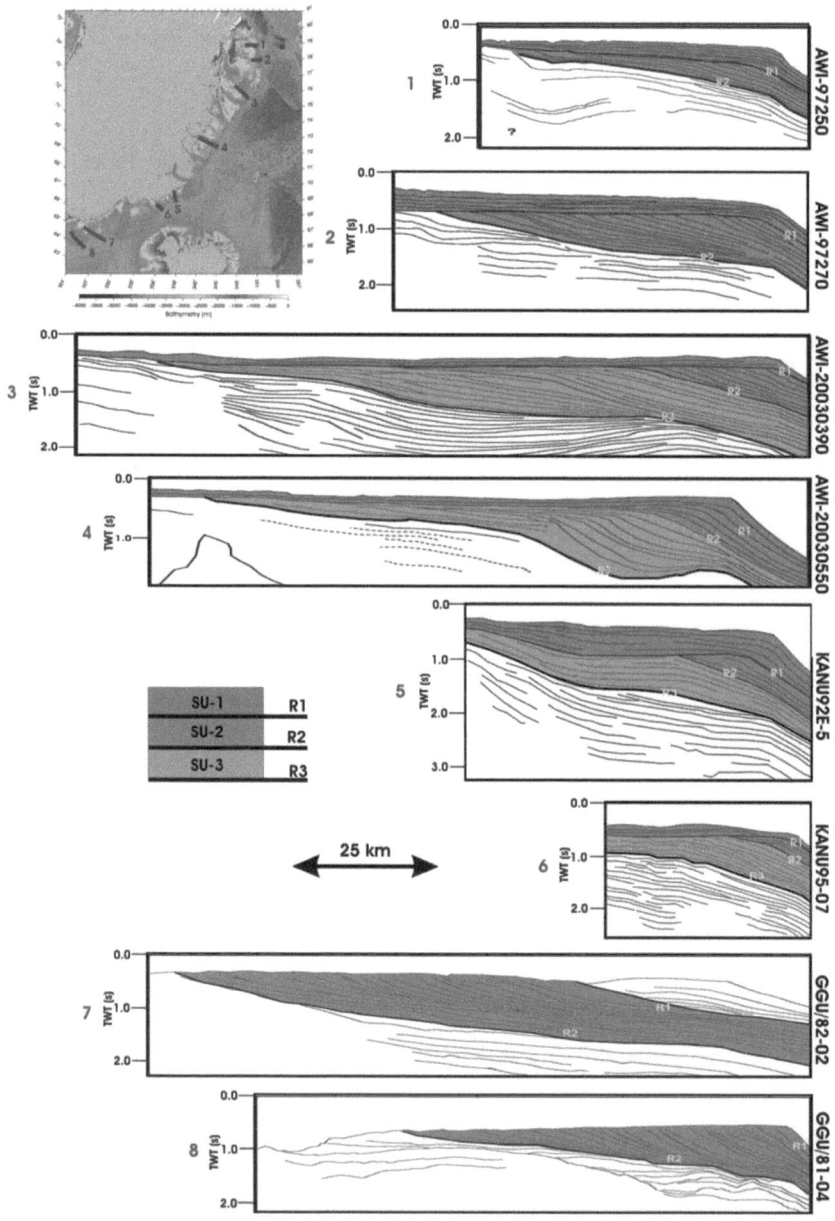

Figure 6.6: Development of the prograding sequences. Line drawing of eight seismic profiles running perpendicular to the East Greenland shelf from north to south. The overview map (top left) shows the location of the profiles (highlighted in red) arranged from north to south. SU-1 to SU-3 are seismic units; R1, R2 and R3 are reflectors separating seismic units. Profiles GGUi82-02 and GGUi81-04 were made available from GEUS. Profiles KANU95-07 and KANU92E-5 were made available by StatoilHydro.

The base of unit SU-2, represented by reflector R2 defines the base of the completely truncated prograding clinoforms. Additionally, the unit SU-2 consists of strongly prograding foresets with dip angle increasing from the older to the younger part. In comparison to unit SU-3, aggradational topsets are missing and the prograding clinoforms in unit SU-2 are characterized by sharp truncations on top, at boundary R1. Within unit SU-2 steep foresets up to the shelf edge are visible. The steepest dip angle of all foresets is seen on the outer shelf on the southernmost profile (Fig. 6.6: GGUi/81-04). On all profiles we could distinguish the sharp truncations of the upper parts of the prograding sequences from the combination of truncated horizons, continuous sequences and gentle dips of the foresets in unit SU-2. The highest sediment accumulation of this unit was observed on the two southernmost profiles GGUi/82-02 and GGUi/81-04 over a shelf distance of 70 km. Seen on profile GGUi/82-02 on the eastern end of the shelf, a sediment package exhibits the typical aggradational character. The beginning of the non-prograding sediment bedding on this profile starts approximately 25 km westward of the present shelf edge. The base of this formation is formed by R1 (Fig. 6.6).

- Top unit (SU-1):

Most of the prograding sequences are overlain by a top unit (SU-1), which is characterised by acoustically horizontal layers (Fig. 6.6). The thickness of unit SU-1 differs from line to line. For example profile AWI-97250 shows an increase in sediment thickness of unit SU-1 towards the shelf edge, where as profile AWI-97270, shows a decrease in sediment thickness in the eastern direction (Fig. 6.6). On both of these profiles, however, the sea bottom appears quite irregular. In the shelf region adjacent to the Greenland Basin, the thickness of unit SU-1 can be described as nearly constant. The unit SU-1 on the northern KANUMAS profile (Fig. 6.6: KANU92E-5) is thinner on the inner shelf than on the outer shelf. These two latter profiles show the thickest deposition of unit SU-1 of all imaged west-east profiles. At the two southernmost profiles (Fig. 6.6: GGUi/82-02; GGUi/81-04) unit SU-1 is absent or below the seismic resolution. In general, these southern profiles look completely different compared to the northern ones, as reflector R1 is missing on these profiles, and thus the prograding foresets of SU-2 are truncated above by the sea-floor in this area.

6.6 Interpretation

6.6.1 Sediment accumulation in the basins along the East Greenland margin

For the calculation of sediment thickness maps along the East Greenland margin, the Geological Survey of Denmark and Greenland (GEUS) have made available datasets south of 72°N. The data north of 72°N give information on the sedimentation into the deep basins along the northern East Greenland margin. The GEUS data in the south cover only the shelf and upper slope area. After a depth conversion (velocity of 2 km/s for the sediment package, Larsen et al., 1994) of the finally processed time sections from the GEUS, the data were used to map the total sediment thicknesses along the entire East Greenland margin. However north of 72°N, a detailed analysis of the sediment succession (subdivision of sediments older and younger as middle Miocene) could be made, on the basis of an age correlation from ODP Site 913, located in the deep Greenland Basin (Berger and Jokat, 2008) and our correlation from ODP Site 909 in the deep Molloy Basin. The results from

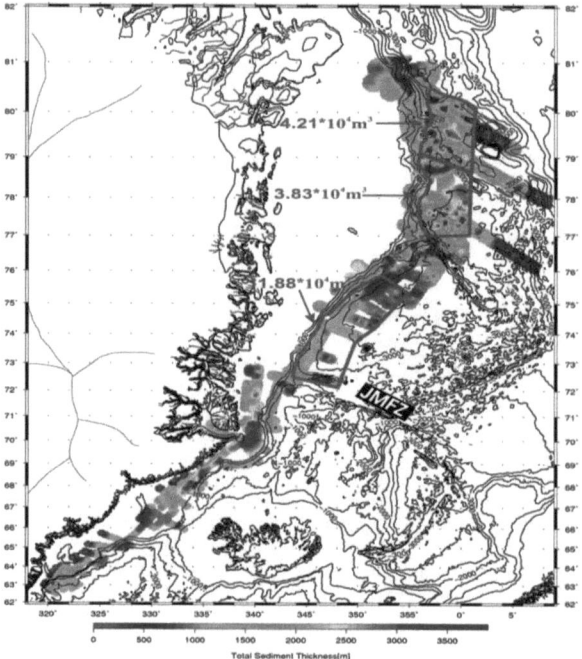

Figure 6.7: A map of total sediment thickness along the East Greenland margin. Bathymetric contours are plotted with a spacing of 500 m. Sediment thicknesses south of 72N published in Larsen et al. (1990), have been incorporated in this map using an average velocity of 2 km/s for the sedimentary units. MB = Molloy Basin, BB = Boreas Basin, GB = Greenland Basin. The red framed regions represent the areas used for the volume calculations.

Berger and Jokat (2008) have also been included into the sediment thickness maps prepared in this study. Berger and Jokat (2008) classified sediments within the Greenland Basin (GB), younger and older than middle Miocene. Therefore, we could join unit GB-2 (Berger and Jokat, 2008) with unit NA-2 and unit GB-1 with unit NA-1. The correlation in the Boreas Basin is based on the interpolation along the north-south profiles 20020540 (Fig. 6.4) and 20020620 (Fig. 6.5). The rough basement structure along the other north-south profiles from the Molloy Basin into the Boreas Basin, prevents the correlation into the southern basin. Different sediment thickness maps are introduced and discussed below, i.e. total sediment thickness (Fig. 6.7), unit I to unit IIIA (Fig. 6.8), unit NA-1 (Fig. 6.9) and unit NA-2 (Fig. 6.10).

Total sediment thickness

The total sediment thickness (Fig. 6.7) was calculated as the depth difference between the acoustic basement and the sea-floor. No continuous increasing or decreasing trends in thickness could not be observed. The thickest sediments (3.3 km) were accumulated in the prolongation of the Scoresby Sund at 70°N, as major glacial wedge (Dowdeswell et al.,

6.6 Interpretation

1997). Some of the material (around 3000 m) of the glacial fan was deposited southwards as far as 67°30'N, probably caused by current-controlled deposition influenced by the East Greenland Current(Rudels et al., 2002).

If we compare the three northern basins between 74°N and 80°N, it is seen that the thickest sediments were accumulated in the Boreas Basin (BB) with 2.7 km (Fig. 6.7). An average of 2.3 km of sediment was deposited in the adjacent northern Molloy Basin (MB), while in the Greenland Basin (GB) a total sediment thickness of around 1 km is observed (Fig. 6.7). Much more representative is the volume of sediments in dependence on the size of the accumulation area. Therefore, volume estimations for the Molloy, Boreas and Greenland

Basin name	Volume ($*10^4 km^3$)	area ($*10^4 km^2$)	volume for an area of $1.0 * 10^4 km^2$
Greenland Basin	11.88	8.68	1.37
Boreas Basin	3.83	2.05	1.87
Molloy Basin	4.21	2.65	1.59
Greenland Basin	7.50	8.68	0.86
Boreas Basin	2.19	2.05	1.07
Molloy Basin	3.29	2.65	1.24
Greenland Basin	4.38	8.68	0.50
Boreas Basin	1.64	2.05	0.80
Molloy Basin	0.92	2.65	0.35

Table 6.1: A table of volume calculations for the Molloy, Boreas and Greenland basins. Gray = total sediments; Blue = sediments younger than middle Miocene and Green = sediments older than middle Miocene.

basins were made (areas are highlighted in figure 6.7). The results of the volume estimation provide for an area (Fig. 6.7) of $8.68*10^4$ km^2 a volume of $11.88*10^4$ km^3 (GB), for $2.05*10^4$ km^2 a volume of $3.83*10^4$ km^3 (BB) and for $2.65*10^4$ km^2 a volume of $4.21*10^4$ km^3 (MB). To compare the total sediment volume, we calculated the volume for a consistent reference level of $1.0*10^4$ km^2 (Tab. 6.1). Thus, we get a total volume of sediment of $1.37*10^4$ km^3 for the GB, $1.87*10^4$ km^3 for the BB and $1.59*10^4$ km^3 for the MB (Tab. 6.1). This means, we have a difference in the sediment deposition of 26.7 % for the GB compared to the BB with the greatest sediment accumulation. The Molloy Basin exhibits 15 % less sediments than in the southern Boreas Basin and 14.8 % more sediment than in the Greenland Basin. So the greatest volume of sediment (of $1.87*10^4$ km^3) is located in the Boreas Basin and the least volume is situated in the Greenland Basin with $1.27*10^4$ km^3.

We explain the lower sediment accumulation in the Greenland Basin in comparison to the two northern basins with, to be due to, the absence of basement highs in the slope region (Berger and Jokat, 2008; Hinz et al., 1987). This structure presumably a 360 km long volcanic basement structure between 77°N and 74°54'N (Berger and Jokat, 2008) prevented continuous sediment transport from the East Greenland shelf to the deep Greenland Basin. North of the GFZ this volcanic structure is not observed, so that sediment material could accumulate constantly. The variations of sediment volumes in both northern basins average only 15% ($1.59*10^4$ km^3 and $1.87*10^4$ km^3) for an area of $1.0*10^4$ km^2 in spite of different times of the opening of these basins. If we assume, the accumulated material in the basins was eroded off the East Greenland shelf, then the erosion was stronger on the shelf adjacent to the younger Molloy Basin than on the shelf area west of the Boreas Basin, for the same time period. Otherwise, a palaeoceanographic model from Ehlers et al. (submitted

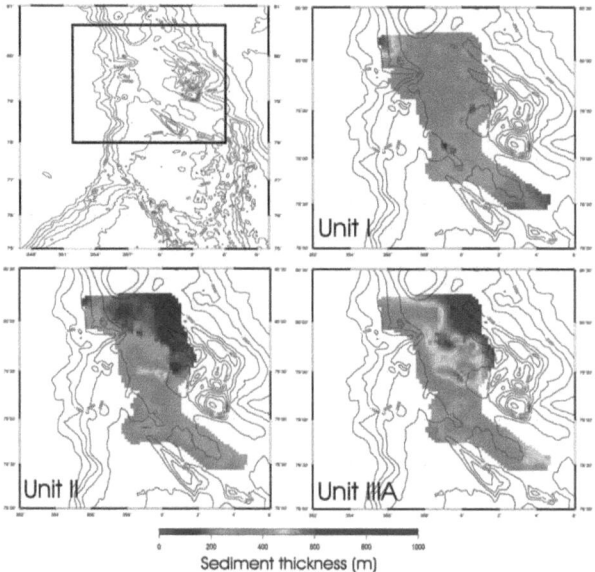

Figure 6.8: Sediment thickness maps for the area of fine bedded sediments (see also figure 6.1). The black box on the top left map indicates the area where calculations of sediment thicknesses for unit I up to unit IIIA were carried out.

2008) shows deep water inflow from the Arctic Ocean towards the northern North Atlantic. Therefore, it is possible that the sediment accumulation in the deep Molloy Basin is also influenced by current-controlled sedimentation as suggested by Knies and Gaina (2008).

Thicknesses of fine bedded sediment units (I-IIIA)

Figure 6.8 shows calculated sediment thickness maps for the area of fine bedded sediments between 78°30'N and 80°30'N (Fig. 6.1). The base of unit IIIB is located in an area of transparent reflection character and thus we could not create a sediment thickness map for this unit. The highest sediment accumulation is seen in unit IIIA, in the northeast region of the Molloy Basin with over 600 m of sediments. In the same area, unit II shows a minimum deposition with less than 200 m of sediment. For both units the surrounding areas show a nearly constant sediment thickness of approximately 400 m. The combined sediment thicknesses of unit II and unit IIIA show a homogeneous sediment distribution of the fine-bedded sediments in a time interval of early/middle Miocene to Pliocene. Unit I images a uniform sediment accumulation of around 400 m. Due to the location of the area (south of the Fram Strait), the refection character and the results from the palaeoceanographic modelling made by Ehlers et al. (submitted 2008) we believe the sediments are current-controlled deposits. Those three units show north of 79°30'N differences in sediment thickness compared to each other, also seen in the sedimentation rates with ∼133m/Ma (unit I), ∼38m/Ma (unit II) and ∼90m/Ma (unit IIIA). The greatest sediment accumulation within unit IIIA is shown close to the deep-water passage between the Arctic Ocean and northern North Atlantic. Before the deep-water exchange begun, northwards flowing North Atlantic deep-water existed (Ehlers et al., submitted 2008). Therefore, we suggest a lot of sediments were transported with the northwards directed current. At this early stage the Fram Strait probably acted as a dam, the current velocity increased and material could be accumulated. The thickness of unit II demonstrates the opposite to unit IIIA. The lowest sediment accumulation is shown north of 79°30'N and probably caused by tectonic. The Fram Strait represents a narrow passage with a limited west-east extent. The further north the more narrow the deep-water passage and the faster the currents in their flow velocity through this gateway. More less sediment will be accumulated with high flow velocities and that can be a reason for the small sediment load in the northern part of unit II. In Pliocene times (unit I) the Fram Strait seems to be wide enough for continuous sediment transport and accumulation by northwards and southwards directed currents.

Thickness of sediments younger and older than middle Miocene

The sediment thickness of unit NA-1 was calculated as the depth difference between the top of the acoustic basement and the lower boundary of unit NA-2 (Fig. 6.9). Unit NA-1 shows the greatest sediment thickness in the northern Boreas Basin with a thickness of approximately 2000 m. The lower part of this unit is characterised by a transparent reflection character most prominent in the Molloy Basin (Fig. 6.3). The sediment thicknesses in the Molloy and Boreas basins vary between 200 and 2000 m. The top of the basement is very rough north of the GFZ, which was caused by ulta-slow spreading of the Molloy and Knipovich ridges (Ehlers and Jokat, 2008). More than twice as much sediments were found (Fig. 6.10) in the deep northern Boreas Basin (2000 m), compared to the deep central Molloy Basin (on average 900 m). These results are also supported by a volume estimations within these basins. For a uniform reference level of $1.0*10^4$ km^2 a sediment volume (Tab. 6.1) of $0.80*10^4$ km^3 for the Boreas Basin and $0.35*10^4$ km^3 for the Molloy Basin (unit NA-1) was calculated. A reason for the different thicknesses in the basins could be the different formation times of the basins. The opening of the Boreas Basin started 38 Ma and 17 Myr later the opening of the Molloy Basin began (Ehlers and Jokat, 2008). This implies, the Molloy Basin is half as old as the Boreas Basin, and explains why there is twice as much sediment in the Boreas Basin compared to the Molloy Basin, provided that the sediment transport was approximately constant over this time period.

The thicknesses of sediments younger than the middle Miocene between 72°N and 80°30'N are mapped in figure 6.10. Both the sediment thickness map and the volume estimations (Tab. 6.1) of these sediments show the greatest accumulation of post-middle Miocene sediments in the Molloy Basin. More than 1000 m of sediments with a volume of $1.24*10^4$ km^3 (area: $1.0*10^4$ km^2) are located in the northernmost basin along the East Greenland margin. However, the basins north and south of the GFZ (Fig. 6.10) feature a similar sediment deposition with around 600 m in the Boreas Basin, and 400 m in the central Greenland Basin. The volume calculation shows a difference of nearly 20% of the sediment volume for an area of $1.0*10^4$ km^2 (Tab. 6.1: GB = $0.86*10^4$ km^3, BB = $1.07*10^4$ km^3) in both basins. Thus, the smallest thickness of sediment in unit NA-2 is situated in the oldest basin of the North Atlantic (GB) with an age of 56 Ma (Talwani and Eldholm, 1977).

A detailed bathymetry map of the East Greenland shelf region (Fig. 6.11) shows the distribution of glacial troughs along the margin. On the shelf, adjacent to the Greenland Basin, between 72°N and 77°N we can find several deep troughs (up to approximately 550 m deep)

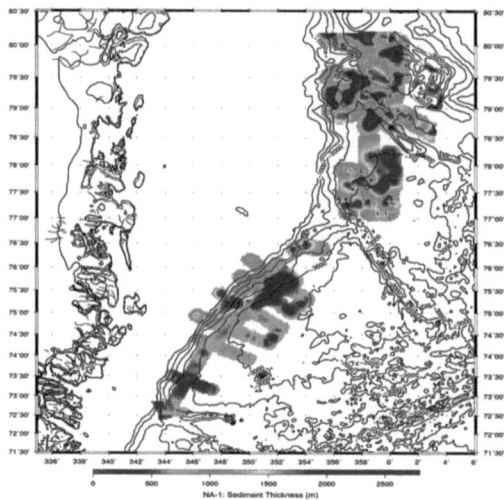

Figure 6.9: A map of sediment thickness for sediments older than middle Miocene (unit NA-1) in the Greenland (GB), Boreas (BB) and Molloy (MB) basins. Bathymetric contours are plotted with a spacing of 500 m.

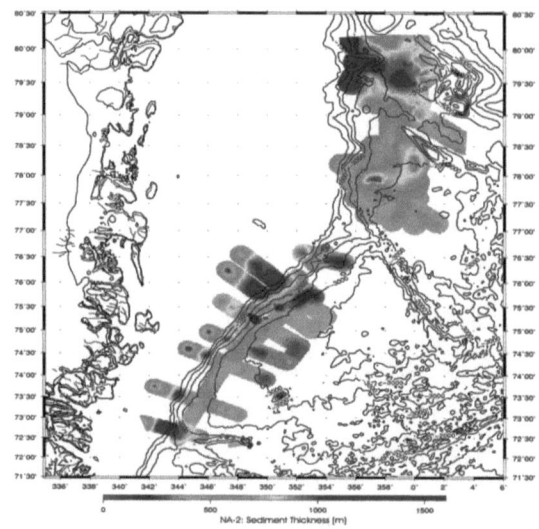

Figure 6.10: A map of sediment thickness for sediments younger than middle Miocene (unit NA-2) in the Greenland (GB), Boreas (BB) and Molloy (MB) basins. Bathymetric contours are plotted with a spacing of 500 m.

in NW-SE direction. We suggest that most of the sediments were transported from the Greenland mainland along these glacially formed channels and accumulated in the prolongation over the shelf edge. The shelf area between 77°N and 81°N appears different to the region south of 77°N, where no glacial troughs with water depths up to 550 m are visible. However, the rough sea-floor topography in approximately 200 m water depth is visible, probably caused by iceberg scours.

Based on the distribution of glacial troughs, we would expect the largest sediment accumulation in the Greenland Basin. However, the prominent volcanic structure mapped by Berger and Jokat (2008) along the East Greenland margin prevents a sediment transport from the shelf into the deep Greenland Basin. This could explain the minor sediment deposits in the Greenland Basin. The largest amount of sediment deposits younger than middle Miocene, are located in the upper slope regions adjacent to the Molloy and Greenland basins. The variations of the sediment deposits in the two northernmost basins are difficult to explain (Fig. 6.10, Tab. 6.1: $1.07*10^4$ km^3 for the BB and $1.24*10^4$ km^3 for the MB). The wider shelf adjacent to the Molloy and Boreas Basin in comparison to the shelf west of the Greenland Basin were caused by the tectonic evolution. During the opening of the Norwegian-Greenland Sea, the plate motion took place in NNW-SSE direction (Lundin and Doré, 2002), and created an asymmetrical decoupling which is apparent. From north to south we see a decrease in the extension of the shelf on the Greenland margin and an increase on the conjugated Norwegian margin (Lundin and Doré, 2002). Because of this, we can not assume that a wider shelf implied a greater mass transport from the East Greenland shelf. Knies and Gaina (2008) suggest that the uplifted northern Barents Sea was ice covered, and that glacial erosion and calving of icebergs along the coastline followed by subsequent transport via Transpolar Drift were the determining processes for the supply of detritus to the Fram Strait. Thus, we suggest the sediment distribution in the Molloy Basin is influenced by current-related sediment transport from the Arctic Ocean and the sediment transport from the Greenland mainland.

6.6.2 Age model of the prograding wedge along the East Greenland margin

The sediments that created the prograding wedges along the East Greenland margin are related to the erosional history and geology of the hinterland as well as the processes of sediment transport and deposition.

The development of a consistent seismostratigraphic model from north to south along the East Greenland shelf was not possible. In order to make an approximate age model, we combined results from age correlations carried out by Larsen et al. (1994) on the south East Greenland margin, Berger and Jokat (2008) on the Greenland shelf (72°N - 77°N) and characteristic interpretations of shelf sediments on the Norwegian margin (Rise et al., 2005).

The stratigraphic correlation between 72°N and 77°N made by Berger and Jokat (2008) provide an age of around 15 Myr for R3 unconformity, which divides aggradational or weak prograding strata from strongly prograding sequences above. The authors suggest rapid changes in sea level and/or glacial erosion by early ice sheets or glaciers along the coast as possible causes for this change. Also drilling results in the southern hemisphere on the glaciated margin of Antarctica support the interpretation of Berger and Jokat (2008) about aggradation and progradation. Cooper et al. (1991) interpret the transition from generally aggrading seismic sequences to prograding sequences at the Ross Sea shelf (Antarctica) as the onset of a large sized grounding ice sheet across the continental shelf. A clear transition from aggrading to prograding strata has not been observed on profiles south of the Scoresby Sund. Especially the two southernmost profiles (Fig. 6.6: GGUi/82-02; GGUi/81-04) show no evidence of aggrading sequences. For this reason, it was not possible to make an age definition for the oldest prograding strata.

The lower unit SU3, described as a mixture of truncated and continuously prograding sequences is well preserved on the shelf between 68°N and 77°N (Fig. 6.6: AWI-20030390, AWI-20030550, KANU92E-5, KANU95-07). The greatest west-east extension of this unit provides profile AWI-20030390 with approximately 80 km. The combination of truncated foresets and partly horizontal horizons within this unit SU-3 are best preserved on this profile (Fig. 6.6). The partly horizontal horizons in the older part of the section could indicate that the advanced glacier did not always reach the shelf break in glacial times and eroded the upper sediments. Otherwise, these horizons can also indicate strong sea level changes. The sea level changes can be caused by the opening of the Fram Strait in middle Miocene times (Ehlers and Jokat (2008); Engen et al., 2008) and the direct deep water connection of the northern North Atlantic to the Arctic Ocean (Kristoffersen, 1990). Winkler et al. (2002) postulate different cooling phases in the middle Miocene, based on the decrease of smectite to illite and clorite ratio at ODP Site 909. They interpret the intensification of the major glaciation in the Northern Hemisphere to have occured around 3.4 to 3.3 Ma, which is synchronous with the age of the dropstones found in the first 250 m.b.s.f. at ODP Site 909 (Myhre et al., 1995). Knies and Gaina (2008) suggest that large-scale glaciation was already developed in the northern Barents Sea during the middle Miocene Climate Transition, around 10-14 million years ago. These findings show that subsequent to an ice-free period during the Miocene Climate Optimum (around 17-15 Ma), glacially eroded materials from the uplifted northern Barents Sea, were transported by iceberg flotillas towards the Fram Strait. The sea level curve from Vail et al. (1977), shows significant fluctuations in Middle Miocene time, suggesting an interaction between climatic changes, sea level changes and the glaciation.

In comparison to the profiles north of the Scoresby Sund, the profiles to the south show approximately more aggradation than progradation (Fig. 6.6: KANU92E-5; KANU95-07) or unit SU-3 is completely missing (Fig. 6.6: GGUi/82-02; GGUi/81-04). The absence of a thick prograding unit especially between 68°N and 69°N, can be explained with:

- The direction of the seismic profiles not coinciding with the axes of the wedge progradation (KANU95-07, KANU92E-5).

- The sediment erosion and deposition by glaciers being too low to build out a strong prograding wedge, for example the north and south of the KANUMAS profiles.

The seismic sequence SU-2 deposited above R2, has the typical characteristic of glaciated margin sequences, truncated by erosional surfaces (Fig. 6.6). The prograding foreset wedge and the preservation of offlap terminations, demonstrate that SU2 is a product of several ice advances towards the shelf edge. The shelf break has prograded seaward during the period of glacial deposition. Ice streams could have migrated laterally, and/or have expanded differentially onto the continental shelf during the last 7 Myr (Nielsen et al., 2005).

Age correlations on the Norwegian margin and the East Faroer margin made by Eidvin et al., 1998; Rise et al., 2005 and Andersen et al., 2000 postulate an age of Pliocene (Norwegian margin) and late Neogene (East Faroer margin) for the beginning of the progradation. The start of the prograding wedge on the South East Greenland margin was dated by Larsen et al. (1994) on the basis of drilling results on the shelf to around 7 Myr. Thus, unit SU-2 on profiles GGUi/82-02 and GGUi/81-04 have an age-range of middle Late Miocene to Pliocene. If unit SU-2 north of the Scoresby Sund has the same age as SU-2 on the southern East Greenland margin it would mean that the onset of glaciation occurs at the same time

6.6 Interpretation

along the entire East Greenland margin. However, this is a very speculative scenario due to the mentioned difficulties in correlating the seismic units along the margin. Non-prograding sediment deposits occur on profile GGUi/82-02 above reflector R1. This profile is located near to the Scoresby Sund and the aggrading well-defined sediment package occurs in the upper slope region. At a water depth of around 600 m, the East Greenland Current is flowing in the southern direction and is probably transporting all the material accumulated in the prolongation of the Scoresby Sund towards south. Therefore, we believe that the sediment accumulation above the prograding sequence on profile GGUi/82-02 is a product of current-controlled sediments.

The upper part of the wedge is topped by an unconformity R1 and overlain by top unit SU-3, which displays horizontal reflection events parallel to the present sea-floor horizon. On many prograding wedges along the European margins, e.g. Sula Sgeir Fan, East Faroer wedge and mid-Norway wedge, this unconformity is interpreted as a glacial unconformity (Dahlgren et al. 2005; Nielsen et al., 2005). On the mid-Norway prograding wedge the glacial unconformity, called Naust B, has an age of 0.35 Ma (Dahlgren et al., 2002) and on the Sula Sgeir Fan and East Faroer Wedge it has been dated to about 0.44 Ma (Dahlgren et al., 2005; Stoker et al., 1995). We suggest that the erosional unconformity on the conjugated Greenland margin is of similar Pleistocene age. The topsets on the South East Greenland margin (Fig. 6.6: GGUi/82-02; GGUi/-81-04) are completely eroded, most likely during the last glacial erosion. During erosion in glacial cycles, the accumulation of glacial sediment enhanced a landward dip of the sea-floor profile. The continental shelf shows a shallowing of the outer shelf and an overdeepening of the inner sector (De Santis et al., 1999), seen on profile GGUi/81-04 and on profile AWI-20030350 published in Berger and Jokat (2008). The sea-floor topography on the two northern profiles (Fig. 6.6: AWI-97250 and AWI-97270) present small irregularities. We suggest, the base of the glacier reached the sea-floor and transported all the material towards the shelf edge. Therefore, the irregularities on top of the sea-floor will be interpreted as iceberg scours, which are attributes of a tremendous ice covering during the last glacial maximum.

6.6.3 Ice movements

If we compare the shelf profiles along the East Greenland margin, we can see regional differences. North of the Scoresby Sund we find the strong development of a prograding wedge up to 77°N with a well-defined boundary (R3) and aggrading sequences below. The detailed bathymetry of the East Greenland shelf (Fig. 11) shows differences between the region south of the Scoresby Sund compared to the area north of 72°N. A much wider shelf between 72°N and 81°N and a more gentle dip of the slope suggests more ice-stream related sedimentation also supported by the deep glacial troughs and fjord systems. Whereas a more narrow shelf in the south and steep slope (Fig. 6.11: GGUi/81-04) could support ice-sheet related sedimentation (Clausen et al., 1997). The two northernmost profiles AWI-97250 and AWI-97270 are also located on the wide shelf, but not in the prolongation of glacial troughs according to the present available bathymetry. Most prominent is the rough sea-floor topography along the two northernmost profiles, influenced by tremendous ice covering in the past. These iceberg scours are not observable on profile AWI-20030390 and profile AWI-20030550 located in the prolongation of glacial troughs and fjord systems (Fig. 6.11). Hence, we believe between 70°N and 77°N there was ice-stream related sedimentation, where as north of 77°N as well as south of Scoresby Sund there was more ice-sheet related sedimentation. The increased aggradation (Fig. 6: KANU92E-5, KANU95-07) combined with a steep slope and missing glacial troughs between 67°N and 68°N (Fig. 6.6, 6.11) suggests that the profiles do not coincide with the

Shelf sedimentation along the East Greenland margin

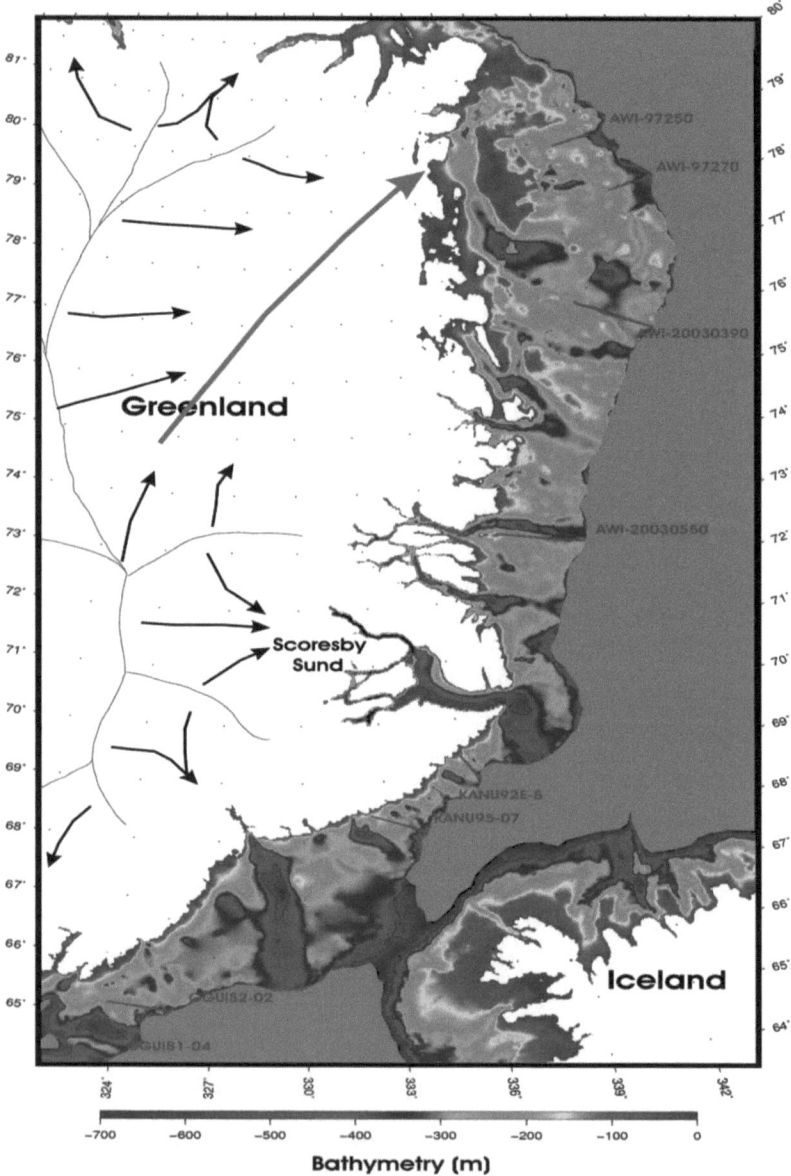

Figure 6.11: Detailed bathymetry map (ETOPO (NGDC, 2006)) of the shelf region along the East Greenland margin. The gray area represents the slope and deep sea parts of the North Atlantic. The red lines on the shelf area show the location of the seismic profiles from figure 6.6. The blue lines on the Greenland mainland demonstrate the ice borderlines, and the black arrows represent the different ice flow directions of the Greenland ice sheet. The green arrow in northeast direction represents the Northeast Greenland ice stream determined by Bamber et al. (2000).

axes of the wedge progradation, erosion and deposition of the glaciers, which is lower than in other regions.
Additionally, in figure 11 we have mapped the ice borderlines (blue lines) of the Greenland inland ice adopt from Bamber et al. (2000). The black arrows demonstrate the directions, where the velocities of the ice streams increase and at the blue lines the ice flow velocity is nearly zero. These calculations are based on a new digital elevation model (DEM), accumulation rates and an existing ice-thickness grid, using a fully two-dimensional finite-difference scheme by Bamber et al. (2000). The green arrow in the northeast direction represents the NE Greenland ice stream published in Bamber et al. (2000). In general, we can see in the prolongation of the different ice flow directions (black) some glacial troughs on the East Greenland shelf. The deepest (700 m) and widest (70 km) troughs can be found near profile AWI-20030390, between profiles KANU95-07 and GGUi/82-02 and the Scoresby Sund. Close to the KANUMAS profiles (68°N-69°N), the bathymetry exhibits poor glacial troughs and from the inland ice towards the shelf break we can not recognize an increase in the ice stream velocity in a southeast direction. An increase of ice stream velocity is observable towards northeast and southsoutheast. Therefore, we do not believe, that much sediment was transported to the outer shelf in that area. On the other side, it is also possible that the more aggradational character on these profiles is a result of sedimentation of glaciers that did not reach the shelf break, and therefore glacio-fluvial drainage was the main sediment transport mechanism. Another situation we have north of 78°N, where also no deep glacial troughs are visible. There we find no aggradation within the shelf deposits but continuously truncated prograding clinoforms, a thick package of flat topset beds and indications of iceberg scours on the sea-floor (Fig. 6.6). This led to the assumption, that all the material was transported by glaciers at different times, beginning around 7 Ma (Larsen et al., 1994). In the prolongation of the NE ice stream, we find a trough north of 80°N (Fig. 6.11). The present bathymetry data show no glacial troughs at the outer shelf in that area. A reason can be the elevation (Fig. 6.11: between 78°30′N and 80°30′N) within the Danmarkshavn Basin (Haman et al. 2005), which looks like a "plateau" on the Northeast Greenland shelf. This structure prevented a direct sediment transport from the Greenland mainland to the outer shelf. In combination with the results of the ice flow directions, we assume the sediments on the inner shelf were transported in a north-south direction along the western side of the "plateau".
Also we think the changing in the direction of ice movements can be an indication of different ice sheets, but on the other hand to discuss real changes in movement, a 3D dataset is needed. It is likelyhowever, that the Greenland ice cap consisted of different ice sheets with different movement directions, and expand at different times.

6.7 Conclusion

The correlation of drill Site 909 made it possible to develop an age model for the northern North Atlantic. On the basis of this seismostratigraphic correlation, a classification of deep sea sediments into unit NA-1 and unit NA-2 was possible for the Molloy and Boreas basins. These units are confirmed with results in the Greenland Basin further south and have been combined in sediment thicknesses maps and volume estimations north of 72°N. These results have shown the lowest total sediment was accumulated in the Greenland Basin compared with the Molloy and Boreas basins, and can be explained with a volcanic structure in the slope region. Variations in total sediment thicknesses in the Molloy and Boreas basins were caused by a stronger erosion on the shelf adjacent to the younger Molloy Basin, than on the shelf area west of the Boreas Basin, in the same time period. More than twice as much sediment older than middle Miocene, was found in the deep northern Boreas Basin (~2000 m) compared to the deep central Molloy Basin (~900 m). We interpret these differences to be due to the formation of these basins at different times. The greatest amount of glacial sediment was deposited in the Molloy Basin, because of influences by current-controlled sediment transport from the Arctic Ocean.

The analysis of seismic sequences on different profiles along the East Greenland margin have made it possible to divide the shelf sediments of unit NA-2 into sub-units based on the results from other polar regions.

- Prograding unit SU3: middle Miocene (~14 Myr) to middle late Miocene (~7 Myr)

- Prograding unit SU2: middle late Miocene (~7 Myr) to Pliocene

- Topset bed unit SU1: Pleistocene

The classification is valid, presuming that the glaciation started at the same time in north and south of East Greenland. On the shelf between 72°N and 77°N we assume more ice-stream related sedimentation, as suggested by the presence of numerous glacial troughs. The demonstration of Greenland ice sheet flow directions led to the assumption, that we have more ice-sheet related sedimentation north of 78°N. South of the Scoresby Sund up to 68°N aggradational sequences are very dominant. The profiles in this region are not located in the prolongation of ice streams of glacial troughs. We believe, however, the glaciers did not reach the shelf break and glacio-fluvial drainage was the main sediment transport mechanism.

6.8 Acknowledgements

We are grateful to the captain and crew of the RV "Polarstern". This research was partly funded by StatoilHydro and the Deutsche Forschungsgesellschaft. We would like to thank Dr. Wolfram Geissler and Dr. Etienne Wildebour Schut for processing the AWI data shown in this study. The authors thank the Geological survey of Denmark and Greenland and the KANUMAS partners for permission to publish interpreted seismic profiles along the Southeast Greenland margin. Furthermore, great thanks is given to John Alexander Cook for improving the English of the manuscript.

6.9 References

Andersen, M.S., Nielsen, T., Sorensen, A.B., Boldreel, O.L., Kuijpers, A., (2000). *Cenozoic sediment distribution and tectonic movements in the Faroe region.* Global and Planetary Change 24 (3-4), 239-259.

Backmann, J., Moran, K., McInroy, D. and the IODP Exp. 302 Scientists, (2005). *IODP Expedition 302, Arctic Coring Expedition (ACEX). A first look at the Cenozoic palaeoceanography of the central Arctic Ocean.* Sci. Drilling. 1, 12-17.

Bamber, J.L., Hardy, R.J., Joughin, I., (2000). *An analysis of balance velocities over the Greenland ice sheet and comparison with synthetic aperture radar interferometry.* Journal of Glaciology, Vol 46, No. 152.

Barker, P.F., Camerlenghi, A., (2002). *Glacial history of the Antarctic Peninsula from Pacific margin sediments*. In: Barker, P.F., Camerlenghi, A., Acton, G.D., Ramsay, A.T.S. (Eds.) Proc. ODP, Sci. Results 178, 40 pp.

Berger, D., Jokat, W., (2008). *A seismic study along the East Greenland margin from 72 - 77°N*. Geophysical Journal International, Vol. 174 (2), page: 733-748.

Cande, S.C., Kent, D.V., (1992). *A new geomagnetic polarity time scale for the Late Cretaceous and Cenozoic*. J. Geophys. Res., 97:13917-13951.

Clausen, L., (1997). *A seismic stratigraphy study of the shelf and deep-sea off Southeast Greenland: the Late Neogene and Pleistocene glacial and marine sedimentary succession*. PhD Thesis, University of Copenhagen, 1-105.

Cooper, A.K., Barret, P.J., Hinz, K., Traube, V., Leitchenkov, G., Stagg, H.M.J., (1991). *Cenozoic prograding sequences of the Antarctic continental margin: a record of glacio-eustatic and tectonic events*. Marine Geology 102, 175-213.

Dahlgren, K.I.T., Vorren, T.O., Laberg, J.S., (2002). *The role of grounding-line sediment supply in ice sheet advances and growth on continental shelves: an example from the Mid-Norwegian sector of the Fennoscandian Ice sheet during the Saalian and Weichselian*. Quaternary International 95-96C, 25-33.

Dahlgren, K.I.T., Voren, T.O., Stoker, M.S., Nielsen, T., Nygård, A., Sejrup, H.P., (2005). *Late Cenozoic prograding wedges on the NW European continental margin: their formation and relationship to tectonic and climate*. Marine and Petroleum Geology, Vol. 22, no.9-10, pp. 1089-1110.

De Santis, L., Prato, S., Brancolini, G., Lovo, M., Torelli, L., (1999). *The eastern Ross Sea continental shelf during the Cenozoic: implications for the West Antarctic ice sheet development*. Global and Planetary Change 23, 173-196.

Dowdeswell, J.A., Kenyon, N.H., Laberg, J.S., (1997). *The glacier-influenced Scoresby Sund Fan, East Greenland continental margin: evidence from GLORIA and 3.5 kHz records*. Marine Geology 143, 207-221.

Ehlers, B.-M., Jokat, W. (2008). *Analysis of subsidence in crustal roughness for ultra-slow spreading ridges in the northern North Atlantic and the Arctic Ocean*. Geophysical Journal International, Vol. 177, Issue 2, pp. 451-462.

Ehlers, B.-M., Butzin, M., Grosfeld, K., Jokat, W., (submitted 2008). *A palaeoceanographic modelling study of the Cenozoic northern North Atlantic and the Arctic Ocean*. Global and Planetary Change.

Eldrett, J.S., Harding, I.C., Wilson, P.A., Butler, E., Roberts, A.P., (2007). *Continental ice in Greenland during the Eocene and Oligocene*. Nature, Vol. 446, 176-179.

Engen, Ø., Faleide, J.I., Dyreng, T.K., (2008). *Opening of the Fram Strait gateway: A review of plate tectonic constraints*. Tectonophysics 450, 51-69.

Gradstein, F., Ogg, J., Smith, A. (eds.), (2004). *A geological time scale 2004*. Cambridge University Press, Cambridge, United Kingdom (GBR).

Hamann, N.E., Whittaker, R.C., Stemmerik, L. (2005), *Geological development of the Northeast Greenland Shelf*. In: Doré, A.G. & Vining (2005), B.A. Petroleum Geology: North-West Europe and Global Perspectives-Proceedings of the 6th Petroleum Geology conference, Volume 2, 887-902, Geological Society, London.

Hinz, K., Mutter, J.C., Zehnder, C.M. and Group, N.S., (1987). *Symmetric conjugation of continent-ocean boundary structures along the Norwegian and East Greenland margins*. Marine and Petroleum Geology 3, 166-187.

Jakobsson, M. et al., (2007). *The early Miocene onset of a ventilated circulation regime in the Arctic Ocean*. Nature, Vol. 447, 986-990.

Jokat, W. et al., (2003). *Reports on Polar and Marine Research*. Mar. Geophys. 449, 8-27.

Knies, J., Gaina, C., (2008). *Middle Miocene ice sheet expansion in the Arctic: Views from the Barents Sea*. Geochemistry Geophysics Geosystems, 9, Q02015, doi:10.1029/2007GC001824.

Kristoffersen, Y., (1990). *On the tectonic evolution and paleoceanographic significance of the Fram Strait gateway*. In: Bleil, U. and Thiede, J. (eds.). Geological History of the Polar Oceans: Arctic Versus Antarctic, 63-76. Kluwer Academic Publishers. Printed in the Netherlands.

Larsen, H.C. (1990). *The East Greenland shelf*. In: Grantz, A., Johnson, L. & Sweeny, J.F. (eds): The Arctic Ocean region. The geology of North America L, 185-210. Boulder, Colorado: Geological Society of America.

Larsen, H.C., Saunders, A.D., Clift, P.D., Beget, J., Wei, W., Spezzaferri, S., (1994). *ODP Leg 152 Scientific Party, 1994. Seven Million Years of Glaciation in Greenland*. Science 264, 952-955.

Larsen, L.M., Fitton, J.G., Saunders, A.D., (1999). *Composition of volcanic rocks from the Southeast Greenland margin, Leg 163; major and trace element geochemistry*. Proceedings of the Ocean Drilling Program, Scientific Results, vol. 163, pp.63-75.

Lundin, E., Doré, A.G., (2002). *Mid-Cenozoic post-breakup deformation in the passive margins bordering the Norwegian-Greenland Sea*. Marine and Petroleum Geology, Vol. 19,

79-93.

Moran, K. et al., (2006). *The Cenozoic paleoenvironment of the Arctic Ocean*. Nature 441, 601-605.

Myhre, A. M., Thiede, J., Firth, J. V. and Shipboard Scientific Party, (1995). *Initial Reports: sites 907-913, North Atlantic-Arctic Gateways I*. Proceedings, Initial Reports, Ocean Drilling Program, ODP 151: 926 pp. (Ocean Drilling Program, College Station, TX).

Nielsen, T., De Santis, L., Dahlgren, K.I.T., Kuijpers, A., Laberg, J.S., Nygård, A., Praeg, D., Stoker, M.S., (2005). *A comparison of the NW European glaciated margin with other glaciated margins*. Marine and Petroleum Geology, Vol. 22, no. 9-10, pp.1149-1183.

Planke, S., Alvestad, E., (1999). *Seismic Volcanostratigraphy of the extrusive breakup complexes in the northeast Atlantic: Implications from ODP/DSDP drilling*. Proc. ODP, Sci. Results, 163, 3-16.

Rise, L., Ottesen, D., Berg, K., Lundin, E., (2005). *Large-scale development of the mid-Norwegian margin during the last 3 million years*. Marine and Petroleum Geology 22, 33-44.

Rudels, E., Fahrbach, E., Meincke, J., Budéus, G., Eriksson, P., (2002). *The East Greenland Current and its contribution to the Denmark Strait overflow*. Journal of Marine Science, 59: 1133-1154.

Stemmerik, L. et al., (1993). *Depositional history and petroleum geology of Carboniferous to Cretaceous sediments in the northern part of East Greenland*. In: Vorren, T.O. et al., (Eds.), Arctic Geology and Petroleum Potential. Norwegian Petroleum Society (NPF), Special Publication 2. Elsevier, Amsterdam, pp. 67-87.

Talwani, M. and Eldholm, O., (1977). *Evolution of the Norwegian-Greenland Sea*. Geological Society of America Bulletin 88, 969-999.

Tsikalas, F., Faleide, J.I., Eldholm, O. and Wilson, J., (2005). *Late Mesozoic-Cenozoic structural and Stratigraphic correlations between the conjugate mid-Norway and NE Greenland continental Margins, Geological development of the Northeast Greenland Shelf*. In: Doré, A.G. & Vining, B.A. (2005) Petroleum Geology: North-West Europe and Global Perspectives-Proceedings of the 6th Petroleum Geology conference, Volume 2, 785-801, Geological Society, London.

U.S. Department of Commerce, National Oceanic and Atmospheric Administration, National Geophysical Data Centre (NGDC), 2006. 2-minutes Gridded Global Relief Data (ETOPO2v2), http://www.ngdc.noaa.gov/mgg/iers/06mgg01.html.

Vail, P.R., Todd, R.G., Sangree, J.B., (1977). *Seismic stratigraphy and global changes of sea level, Part 5: Chronostratigraphic Significance of Seismic Reflections*. In: Payton, C.E.,

(Eds). American Association of Petroleum Geologists, no.26, 99-116.

Winkler, A., Wolf-Welling, T.C.W., Stattegger, K., Thiede, J., (2002). *Clay mineral sedimentation in high northern latitude deep-sea basins since the Middle Miocene (ODP Leg 151, NAAG)*. Int. J. Earth Sci. 91, 133-148.

7 Paper 3

Current-controlled sedimentation along the Northeast Greenland margin

Daniela Berger and Wilfried Jokat

Marine Geology (2009)

submitted in original form 2009 March 19

7.1 Abstract

Seismic reflection data between 72°N and 81°N show current-related sedimentation partly linked to ocean circulations in the Northern Hemisphere. For the first time, it is possible to prove the speculations about the deep-water exchange between the Arctic Ocean and northern North Atlantic with seismic reflection data. An around 25 km wide deep sea channel at the southern end of the Fram Strait images contourite deposits west and east of the channel, which points to a deep-water outflow coming from the north (Arctic Ocean) and a northwards directed current (West Spitsbergen Current) coming from the south. The formation of the channel is dated to have an age of 5 Myrs. However, sediments below the channel system indicate contourite accumulated since the opening of the deep-water passage 17-18 Ma. Within the deep Greenland Basin, a very local channel-levee system demonstrates a northwards directed turbidity current with eastwards accumulated sediment mounds, decreasing in their thickness towards the north. Our explanation is an episodic varying ice stream activity triggered by intensification of glacial and interglacial times, which has probably started 5 Ma. Seismostratigraphical results of sediment desposits (e.g. turbidities, slumps and a sediment drift body) observed in the slope area adjacent to the Greenland Basin, indicate a continuous beginning of the formations before 15 Ma (observed within unit GB-1).

Key words: Channel-levee deposits, Current-controlled sedimentation, Northeast Greenland, Ocean circulation.

7.2 Introduction

Plate motions can strongly influence ocean current circulations and therefore also climatic conditions. The separation of the Australian and African continents from Antarctica (Tasmania-Antarctic Passage) contributes partly to the stepwise development of the Antarctic Circumpolar Current (ACC) around Antarctica. After the final opening the Drake Passage enabled the evolution of the eastward flowing of the ACC (Lawver and Gahagen, 2003). This circulation had a tremendous effect in climate changes in the Southern Hemisphere. A following cooling trend with rapid expansion of the Antarctic continental ice sheet is triggered by the opening of the Drake Passage during the early Oligocene (Lawver and Gahagen, 2003). These observations are supported by analyses of drilling samples combined

with several seismic reflection data sets (Miller et al., 1990; Kuvaas & Leitchenkov, 1992; Brancolini et al., 1995; De Santis et al., 2003).
Geological investigations along the Norwegian margin showed that the closing of the Isthmus of Panama had a high influence on the climate and ice formation in the north (Zachos et al., 2001). The Gulf Current has been diffracted towards north, resulting in more rainfall/snowfall and ice concentration in the High Arctic. Several sediment core analyses and geophysical data confirm the assumptions to an onset of a glacial stage in Pliocene times ($\sim$3 Ma) along the Norwegian margin (Forsberg et al., 1999; Rise et al., 2005). Until recently, it is still unclear if the former opening of the Fram Strait (Fig. 7.1) in middle Miocene times (Jakobsson et al., 2007; Ehlers and Jokat, 2008) had also influence on the Northern Hemisphere climate. The first deep-water exchange between the Arctic Ocean and North Atlantic took place at 17-18 Ma (Jakobsson et al., 2007; Ehlers and Jokat, 2008). A connection between climate changes, the onset of glaciation of the Northern Hemisphere and the opening of the deep-water gateway (Fram Strait) is presently not fully investigated. Missing comprehensive drill site information, seismic surveys and oceanographic studies make it difficult to provide a complete description of the consequences of this gateway opening.
Due to the sparse geoscientific database in the Arctic a lot of speculations about the onset of glaciation exist. They vary from Pliocene (Stemmerik et al., 1993) to late Eocene (Eldrett et al., 2007). Berger and Jokat (2008) correlated age information along seismic lines from the deep sea (ODP Site 913) to the Northeast Greenland shelf to provide an estimate for the beginning of the Greenlandic ice formation. Their interpretation based on seismic reflector geometry is that the onset of glaciation happened in middle Miocene times. This is the same time period where the deep-water exchange between the Arctic Ocean and North Atlantic was established. Eldrett et al. (2007) actually believe in massive ice formations in Greenland already during late Eocene, based on a biostratigraphic analysis of dinoflagellate cysts on site 913 in the deep Greenland Sea.
The seismic data sets acquired since 2002 between 70°N and 81°N provide the possibility to analyse the sediment structure also in terms of current activity near the Fram Strait, which might be an indicator for glacial/interglacial periods. In general, identified channel structures with accompanied sediment levee complexes can point to current activities and their flow directions. Additionally to bottom-current processes along a margin also down-slope sediment deposits are common. Due to sediment overloading at the shelf break e.g. during glacial times, an instability of the continental slope and slump generation can be triggered. Especially for sediments in upper, steep slope regions mass transport by gravity-driven processes plays an important role. Large volumes of sediment were transported from the slope into the deep basins. Numerous, turbidity currents cut into the lower slope and rise on other parts of continental margins i.e. Antarctica and northwestern Sea of Okhotsk (Escutia et al., 2003; Wong et al., 2003; Dowdeswell et al., 2004).
This study concentrates on sediment structures in the region north of the Jan Mayen Fracture zone up to the northern part of the Molly Basin (Fig. 7.1). With the application of the seismostratigraphic concepts for the Greenland Basin (Berger and Jokat, submitted 2008), the Boreas and Molloy basins (Berger and Jokat, submitted 2008) we investigate different current-controlled sediment deposits and where possible we try to link the results to climate changes and their consequences.

7.2 Introduction

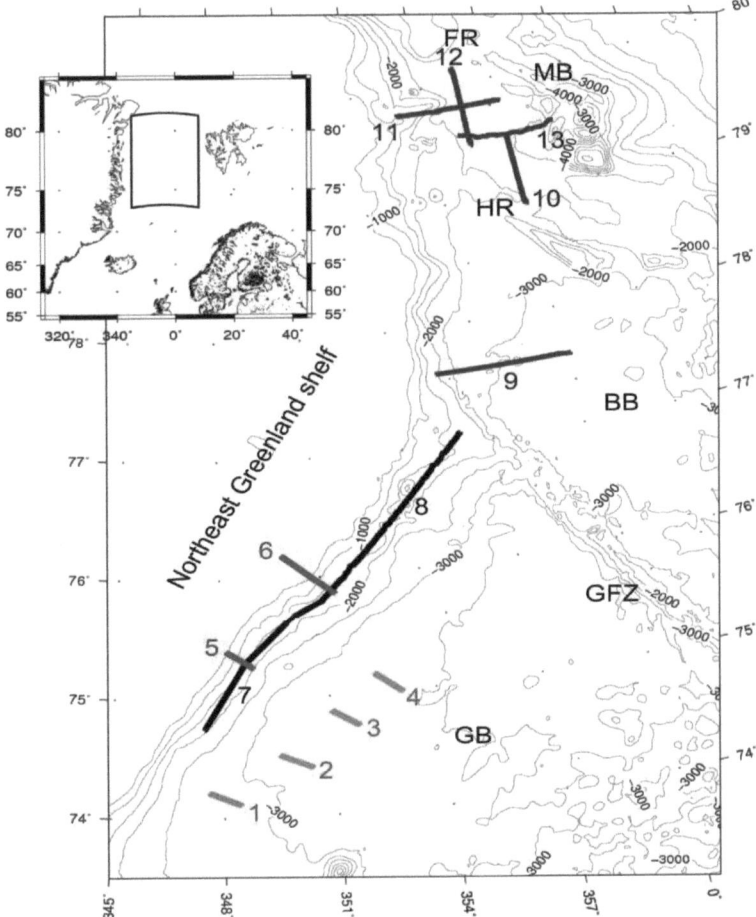

Figure 7.1: Location of the study area on the Northeast Greenland margin. FS = Fram Strait, HR = Hovgård Ridge, GB = Greenland Basin, BB = Boreas Basin, MB = Molloy Basin, GFZ = Greenland Fracture zone. Green profiles (1 - 4) show observed deep sea channel structures (Fig. 7.2), red profiles (5, 6) show gravity-driven sediment structures in the upper slope region (Fig. 7.3), black profiles (7, 8) show drift structures in the slope area (Fig. 7.4), blue profiles (9 - 13) show current influenced sediment structures (Fig. 7.6).

7.3 Oceanographic setting

The North Atlantic-Arctic Gateway is crossing the Greenland-Scotland Ridge in the South and the narrow deep-water passage (Fram Strait) between the continental margins of East Greenland and Svalbard in the North (Fig. 7.1). The Fram Strait, is channelling the flow of surface and deep waters between the Arctic and North Atlantic. It represents the most recent stage in the plate tectonic development of the Atlantic Ocean from a zonal (much like the modern Pacific and Indian oceans) to a meridional ocean basin allowing the deep-water exchange from both polar hydrospheres (Myhre et al., 1995). This passage is located between 78°N and 82°N and the opening of this gateway started some 20 Ma as shallow water connection. The deep water exchange started 17-18 Ma in middle Miocene times (Jakobsson et al., 2007; Ehlers and Jokat, 2008). Today this passage has a width of around 200 km and is limited in the west by the Northeast Greenland shelf and in the east by the Yermak Plateau. (Fig. 7.1)

The ocean water circulations can be divided into surface waters and bottom waters. The surface water current systems of the Norwegian-Greenland Sea include the influx of warm and relatively high-salinity waters via the North Atlantic Drift, which continues its northward flow as the Norwegian Current, and the outflow of cold and low-salinity waters via the East Greenland Current (EGC). The northward current continues from the Norwegian Current along the Norwegian margin into the West Spitsbergen Current along the western margin of Svalbard before entering the Arctic Ocean and dipping under the Arctic sea-ice cover (Myhre et al., 1995). Before this current reaches the Fram Strait it splits into three components (Rudels et al., 2002). One component flows in the eastern direction north of Svalbard along the southern Yermak Plateau. One branch re-circulates north of the Greenland Fracture zone and the third branch transports water northwards and enters the Arctic Ocean through the eastern Fram Strait (Rudels et al., 2002; Ehlers et al., submitted 2008). Within the Arctic, the warm water mass mixes with low salinity and cold surface waters, sinks, and flows as an intermediate water mass counterclockwise before being exported out of the Arctic Ocean via the Fram Strait along the East Greenland margin (Myhre et al., 1995).

Ehlers et al. (submitted 2008) modeled the ocean circulations for four time slices 45 Ma, 20 Ma, 15 Ma and today, based on palaeobathymetric reconstruction data for the northern North Atlantic and the Arctic Ocean. These models show significant changes in ocean mass, heat and salt transport regarding to the bathymetric evolution according to the different seafloor topography.. Their results indicate that the Norwegian and West Spitsbergen currents are today in water depths above 1125 m. Below, 1125 m the northward flow of Atlantic Water ceases south of 75°N at the Barents Sea margin. Some 15 Ma, neither evidences for northward flowing of Atlantic deep-water nor evidences of re-circulating water masses towards the East Greenland margin are shown in their model. The EGC, which transports Arctic Ocean water southwards, exists above 500 m (Ehlers et al., submitted 2008). The current follows the bathymetry of the East Greenland shelf to the Greenland-Faeroe Ridge and can be described as surface water current mixed with re-circulating Atlantic Water.

In general, deep-water formations can have a strong influence on sedimentation in the deep sea. Until now, bottom current deposits in the northern North Atlantic are not well-studied. Deposits controlled by deep-water bottom currents (contour currents) resulting from thermohaline circulation in the oceans form accumulations known as contourite drifts (Faugres et al., 1999). The most extensively studied area of submarine channels is located in the Norwegian-Greenland Sea and occurs on the Northeast Greenland margin between 72°N and 75°N (Fig. 7.1). Several expeditions have been carried out to investigate the distribution and morphology of channel systems in this region by GLORIA 6.5 kHz side-scan sonar, mutlibeam swath bathymetry and sub-bottom profiling (Mienert et al., 1993; Fahrbach, 2002; Lembke, 2003, Stein, 2008). Present channels systems within the Greenland Basin are shown in Stein (2008). Their origin is still under discussion. The channels can be related to down-slope flow of dense-water and turbidity currents originating from sea-ice formation on the shelf and upper slope (Mienert et al., 1993; Dowdeswell et al., 1996, 2002). Cofaigh et al. (2004), however, favour a formation under glacial conditions and Wilken and Mienert (2006) believe in major glaciations with varying ice stream activity across the outer and inner shelf. We will show and interpret seismic reflection data which image these channel and contourite structures within the deep Norwegian-Greenland Sea.

7.4 Description and interpretation of seismic reflection profiles

Seismic reflection profiling was carried out using an array of eight VLF guns with a local capacity of 3l each. The shots were generated every 15 s and binned into 25 m spaced CDPs. A 600 m and 3000 m streamer were used for data acquisition, and before stacking and time migration a bandpass filter of 15-90 Hz was applied. Figure 7.1 shows the seismic profiles off the East Greenland margin, which are used in this study. The entire seismic network in this area is shown in Berger and Jokat (2008) and Berger and Jokat (submitted 2008). These profiles are located in the Greenland, Boreas and Molloy basins (Fig. 7.1). The age information is adopted from Berger and Jokat (2008) and Berger and Jokat (submitted 2008). Four seismic lines in figure 7.2 show a more or less clear deepening in the seafloor topography along these lines (line 1.: CDP 2600, line2: CDP 7850, line 3: CDP 1400, line 4: CDP 8450). The deepening on these profiles ranges between 23 m and 100 m and the width ranges between 0.8 km and 3 km. These seafloor undulations point to a local deep-sea channel within the Greenland Basin. The three southernmost west-east orientated lines (Figs. 7.1, 7.2) show east of the north-south oriented channel system an elongated well stratified sediment deposit, best preserved on line 1 (Fig. 7.2: line 1 east of CDP 2600), which we interpret as channel-related drift package. In general, it seems to be that the evolution of the elongated drift decreases in the northern direction as well as the width of the channel. On profile 4 (Fig. 7.2), the channel structure is scarcely developed and an elongated drift is not visible. The quite small depression between CDP 8300 and 8450 in comparison to the surrounding seafloor topography suggests a channel system on this profile. North of 75.3°N, no indications of channel-related sedimentation could be observed in the Greenland Basin. The reflection character changes from laminated sediments in the upper ~400 ms (TWT) to transparent sediment succession below (Fig. 7.2: lines 1-4). The transparent sedimentary strata, is visible in the lower part of unit GB-2 and the entire unit will be interpreted by Berger and Jokat (2008) and defined as middle Miocene age.

Along the East Greenland continental slope we found small irregularities between continuous down-slope reflectors, which result in weak interruptions of the reflectors (Fig. 7.3: line 5 CDP 5100 (2500 ms) and CDP 5600 (2250 ms), line 6 CDP 4500 (2250 ms) and CDP 5600 (2100 ms)). We interpret these structures as turbidity current sedimentation, which are typical for down-slope gravity-driven processes in slope areas (Faugres et al., 1999). Both profiles in figure 7.3 show evidences for turbidity currents, both in unit GB-1 and GB-2. However, the sediment structure in unit GB-1 looks a bit more chaotic than in GB-2, probably caused by more extensive down-slope transportation as well as due to the presence of more coarse grained material.

The along-slope profiles in figure 7.4 (line 7 and 8) show ripples along the profiles developed

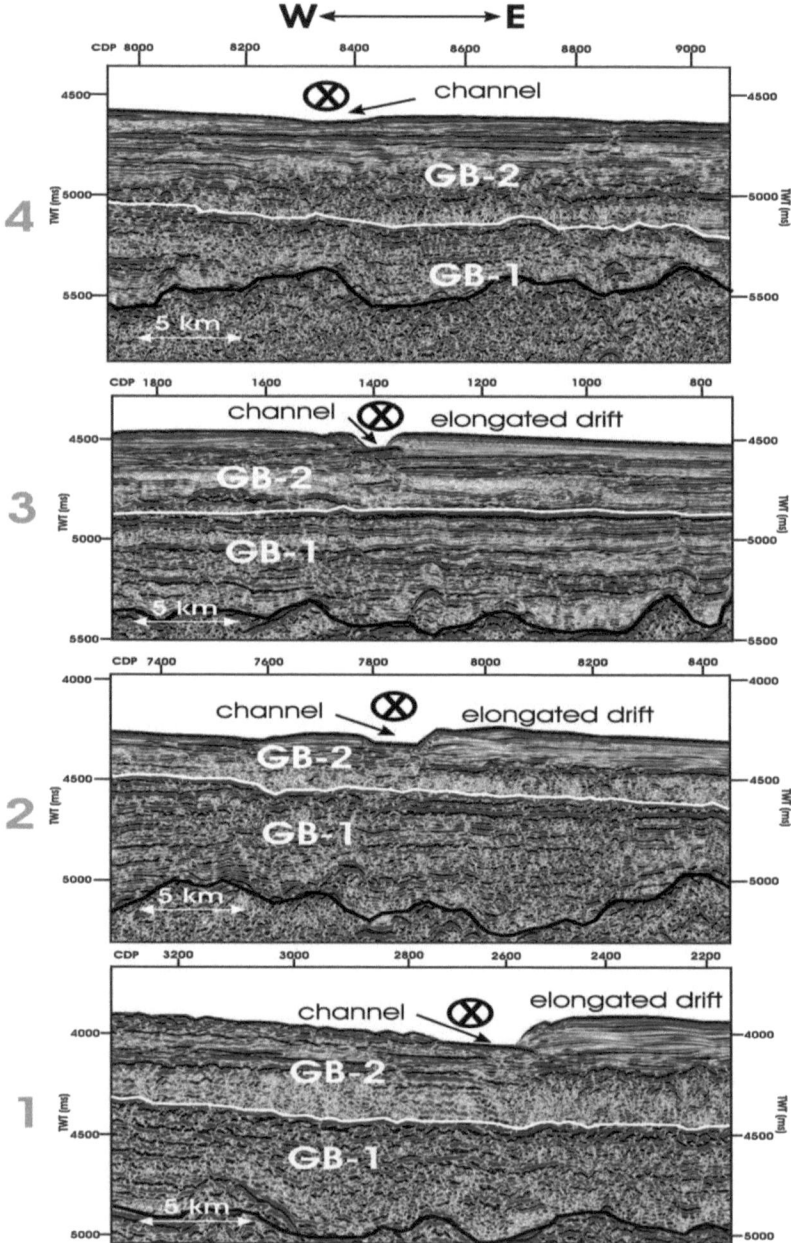

Figure 7.2: Parts of seismic reflection profiles located in the Greenland Basin (for location see also figure 7.1). All profiles show deep sea channel structures and on profiles 1 up to 3 an elongated sediment mount could be observed. The black lines in the seismic data represent the top of acoustic basement. Units GB-1 (Tertiary - middle Miocene) and GB-2 (middle Miocene - present) have been adopted from Berger and Jokat (2008).

7.4 Description and interpretation of seismic reflection profiles

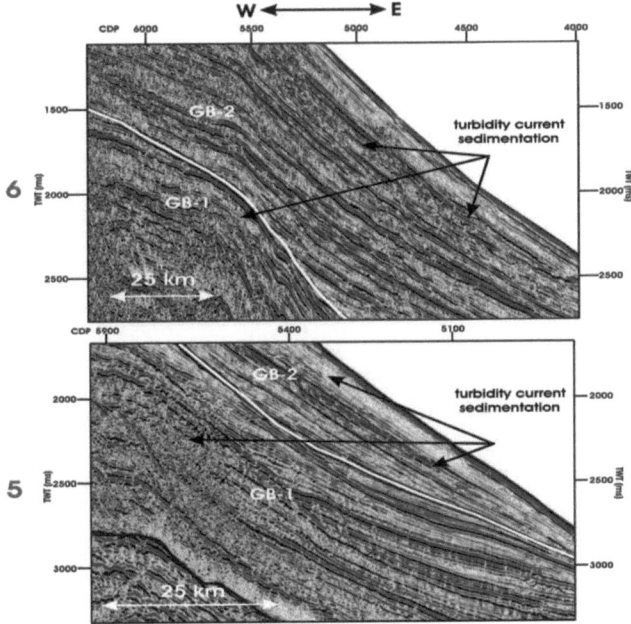

Figure 7.3: Parts of seismic reflection profiles located in the slope region of the Greenland Basin (for location see also figure 7.1). The profiles show downslope structures which point to turbidity currents. The black lines in the seismic data represent the top of acoustic basement. Units GB-1 (Tertiary - middle Miocene) and GB-2 (middle Miocene - present) have been adopted from Berger and Jokat (2008).

at the seafloor. We assume that these seafloor undulations can be interpreted as turbidity channels down the slope. Also the Parasound data show turbidity activity (Matthiesen et al., 2003; Stein, 2006) in that region. On line 7 (Fig. 7.4) a prominent sediment body (CDP 1900) with a width of around 26 km and a high of 0.5 s (TWT) is visible. The base of this structure (CDP 1900 / 3500 ms) is situated within unit GB-1 and the top reflector represents the boundary to unit GB-2 above. That means the initiation of this drift body took place before the massive onset of glaciation on the Northeast Greenland shelf in middle Miocene times (Berger and Jokat, 2008). The along slope profiles (Fig. 7.4) show a chaotic to hyperbolic, occasionally wavy facies with subparallel reflections. The chaotic sediment facies combined with the base of the sediment drift on profile 7 (Fig. 7.4: CDP 1900) suggest to us that the deposit is a result of intensive down-slope activity on the Northeast Greenland slope also supported by glacial troughs in this area on the continental shelf.
North of the Greenland Fracture zone (GFZ) in the Boreas and Molloy basins we find different sedimentation conditions compared to the basin south of the GFZ, as discussed by Berger and Jokat (submitted 2008). In the deep Molloy Basin in a water depth of more than 2500 m we can observe parallel to sub-parallel and well-stratified reflections with high amplitudes (Fig. 7.5; line 10: CDP 400-2700 (3800-5000 ms), line 11: CDP 1-1500 (3800-4800 ms), line 12: CDP 500-3000 (3800-4800 ms). This area seems to be very local and can be limited in the western direction (Fig. 7.1: gray shaded area). The transition from the well-stratified strata to a chaotic sediment accumulation is clearly visible (see also Berger and Jokat, submitted 2008) in profiles 11 (CDP 1-1500 (below 4800 ms) and 12 CDP 500-3000

Current-controlled sediments along the Northeast Greenland margin

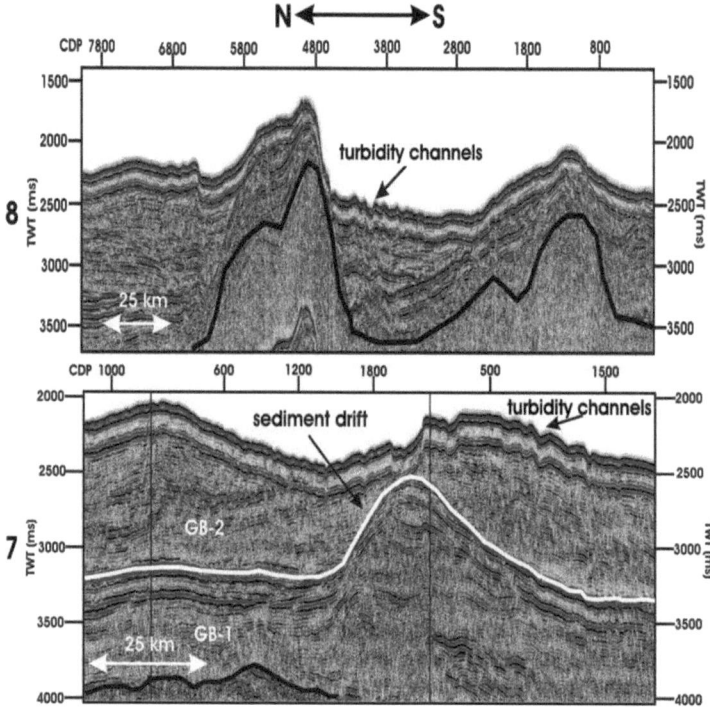

Figure 7.4: Parts of seismic reflection profiles located along the slope region of the Greenland Basin (for location see also figure 7.1). At the top of the seabottom small elevations are visible which point to mudwave activity caused by turbidity currents down the slope. The black lines in the seismic data represent the top of acoustic basement. Units GB-1 (Tertiary - middle Miocene) and GB-2 (middle Miocene - present) have been adopted from Berger and Jokat (2008).

(below 4800 ms) (Fig. 7.5). In general, the changing of the sedimentation character is visible on all west-east profiles in the Molloy Basin. The profile in the southern Boreas Basin (Fig. 7.5: line 9), however, demonstrates a completely different reflection pattern in comparison to the profiles located in the deep Molloy Basin (Fig. 7.5; lines 10 - 12; Fig. 7.6; line 13). Other W-E orientated profiles in the Boreas Basin show the same pattern. On these profiles we find only one well-defined reflector observed in unit NA-2 and unit NA-1 is characterized as complete transparent. A well-laminated sediment supply like in the Molloy Basin could not be recognized on the seismic profiles south of the Hovgård Ridge. The fine-bedded sediments north of 78°30'N (Fig. 7.1) seems to be a result of current-controlled sedimentation, and will hence be interpreted as contourites, which are controlled by deep-water bottom currents resulting from thermohaline circulation in the oceans. These circulations are established by Rudels et al. (2002) in this area. Additionally, we assume the various sedimentation strata point to different sedimentation velocities. The transparent reflection character could be a product of coarse grained material whereas the thin-bedded sediment section consists of fine grained material. Probably the coarse grained sediments have been transported by currents with a higher transport velocity than the fine grained sediments above. On the north-south profile 10 (Fig. 7.5) between CDP 2000 and CDP 2400, we observe a small depression at

106

7.4 Description and interpretation of seismic reflection profiles

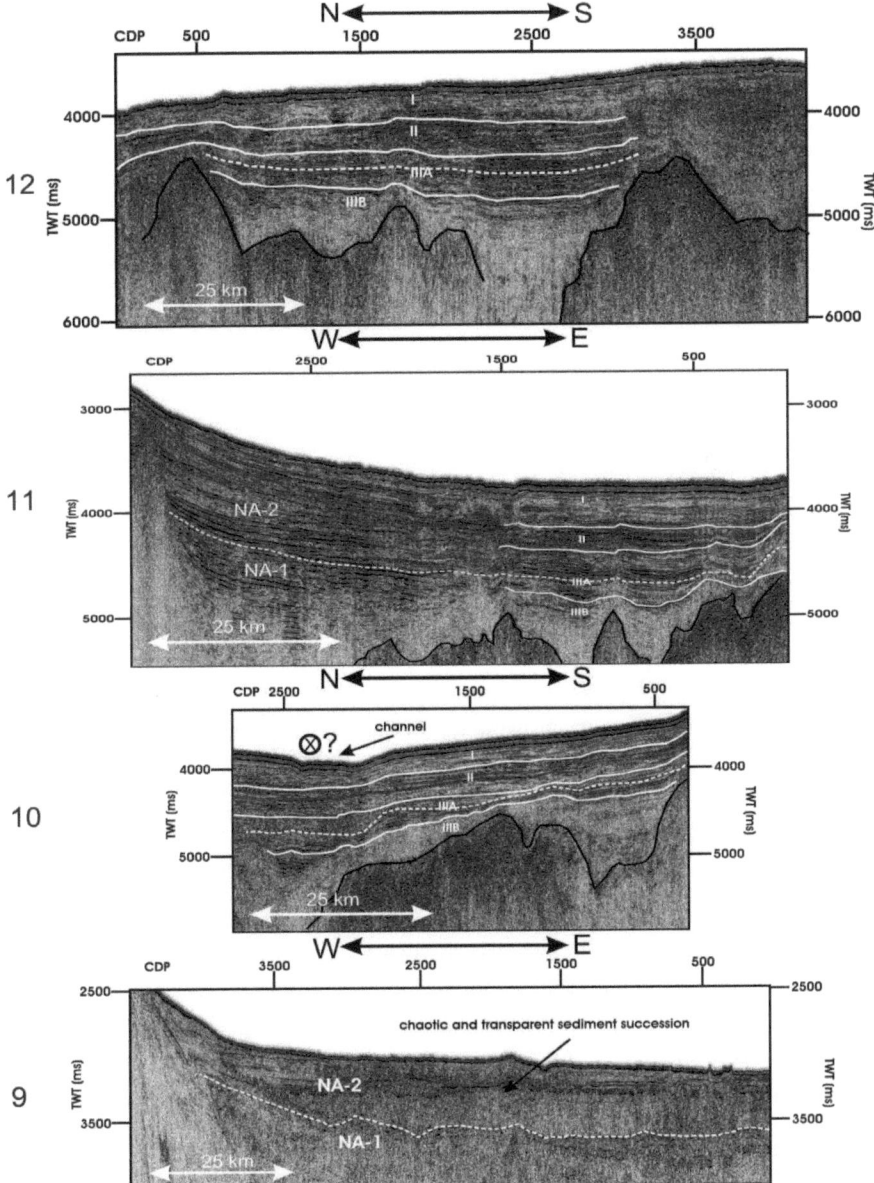

Figure 7.5: Parts of seismic reflection profiles published in Berger and Jokat (submitted 2008). For locations see figure 7.1 (Boreas Basin: line 10; Molloy Basin: lines 11-13). The subdivision of the fine-bedded sediments into four units (based on drilling results from ODP site 909) was adopted from Berger and Jokat (submitted 2008). Unit I = Pliocene to Quaternary, Unit II = Miocene to Pliocene, Unit IIIA = Oligocene to Miocene, Unit IIIB = older than Oligocene. The dotted line represents the boundary between unit NA-1 (older than middle Miocene) and unit NA-2 (younger than middle Miocene). The black lines shows the top of acoustic basement.

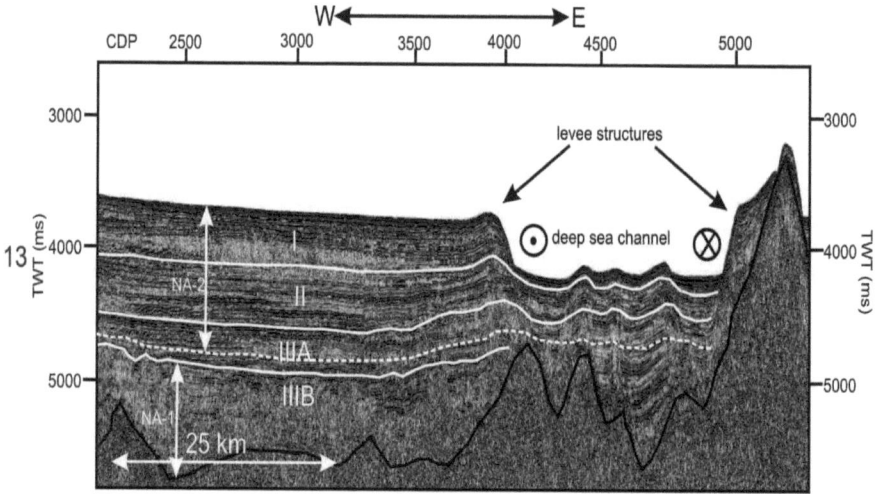

Figure 7.6: Part of a seismic reflection profile located in the central part of the Molloy Basin (Fig. 7.5: line 13). The subdivision of the fine-bedded sediments into four units (based on drilling results from ODP site 909) was made on the basis of Berger and Jokat (submitted 2008). Unit I = Pliocene to Quaternary, Unit II = Miocene to Pliocene, Unit IIIA = Oligocene to Miocene, Unit IIIB = older than Oligocene. The dotted line represents the boundary between unit NA-1 (older than middle Miocene) and unit NA-2 (younger than middle Miocene). The black line shows the top of acoustic basement.

the seafloor. At deeper reflectors the topography is more flat. We, therefore, interpret this depression as a very young channel system. No channel systems and continuously sediment accumulation could be observed in the Boreas Basin. Therefore, we think the current velocity of deep-water ocean circulations was quite high and prevented a building of typical current-related sediment structures. The profile imaged in figure 7.6 crosses the deep Molloy Basin at 79°30'N. From west to east, the seafloor deepens at CDP 4000 and rises at CDP 5100 and the depth of the seafloor ranges between 2800 m and 3150 m. We interpret this structure as a channel system similar to the channel identified in the deep Greenland Basin, but much wider in the west-east extension with around 26 km. At CDP 4000 we observe a fine laminated sediment mound interpreted as sediment levee complex. East of the levee the seafloor deepens from 3700 ms to 4200 ms. At a depth of 4200 ms, three shallow undulations are visible before the seafloor shallows again at the eastern end of this profile (CDP 5150). This structure suggests that sediments have been and are still being transported through the Fram Strait accumulating in the deep Molloy Basin. A striking reflector in a depth of around 4800 ms (Fig. 7.6) marks the base of unit NA-I which is interpreted as onset of glaciation (Berger and Jokat, submitted 2008). Furthermore, Berger and Jokat (submitted 2008) have made an age classification of the thin-bedded sediments located in the deep Molloy Basin (Fig. 7.5). These interpreted age units (I = Pliocene to Quaternary, II = Pliocene to Miocene, IIIA = Miocene, IIB = Miocene to upper upper Oligocene) we have correlated along the profiles 10, 11, 12 (Fig. 7.5) and profile 13 (Fig. 7.6), which gives us the possibility to show the variation of the sediment accumulation to different times. The results show the base of the sediment levee complex on profile 13 (Fig. 7.6: CDP 4000 / 4000 ms) fit with the base of unit I which has an age of Pliocene. Therefore, we suggest the formation of the levee structure started in Pliocene times.

7.5 Discussion

7.5.1 Slope sedimentation

The sedimentation on continental slopes is mainly influenced by interaction between gravitational down-slope and current-controlled alongslope transport mechanisms. These are large-scale bedforms found in ocean basins worldwide, occurring from continental slopes to abyssal plains (Howe et al., 1998). One of the major difficulties encountered at present in the interpretation of seismic profiles across continental margins is the differentiation between turbidite and contourite deposits and their associated facies (Faugéres et al., 1999). Between 75°N and 77°N in the slope region of the East Greenland margin we could recognize mudwave structures on top of the seafloor and in deeper levels both slump and drift structures. The interpreted ripple structures on top of the seafloor along the slope-parallel profiles (Fig. 7.4: lines 7, 8) represent flow channels build up by turbidity currents or debris flows of the downslope sediment transport. The wave form morphology can be traced down into unit GB-1 (Fig. 7.4: line 7), which points to an activation of the process of turbidity formation before 15 Ma (Berger and Jokat, 2008). In figure 7.3, we see the structure of the sedimentation along these flow paths, which is described as mixture of continuous horizontal reflectors with sub-horizontal reflections and reduced amplitudes. Therefore, we assume that the ripples are fed predominantly by localised fine-grained turbidite-derived material laterally transferred by weak bottom current activity. This characterisation is also present in unit GB-1, which confirms the assumption of down-slope transportation before middle Miocene times. These turbidity deposits and turbidity channels have been observed only in the slope region of the Greenland Basin (Fig. 7.7).

Further north, evidences for turbidity activity have not been detected, probably caused by more less down-slope activity. Otherwise the data are limited in the western direction and therefore the shelf and upper slope areas are not well imaged. Howe et al. (1998) suppose on the northwest Weddell Sea that lower sea levels during glacial episodes will help to increase turbidity current frequency and sediment accumulation. Therefore, we believe in glacially triggered processes for turbidite sediment deposition. The absence of turbidity deposits and down-slope channels north of 77°N, as well as less glacial troughs in the shelf region (Berger and Jokat, submitted 2008), suggest minor down-slope sediment transportation. Additionally, on line 7 (Fig. 7.4) we find a prominent sediment drift body with its base within unit GB-1. This slope structure seems to be linked to an age of older than 15 Ma.

A first appearance of mudwave structures or turbidity channels in the seismic record on the northwest Weddell Sea margin by Howe et al. (1998) is interpreted to mark the opening of a deep-water pathway in the Southern Hemisphere. The postulated age of the opening of the Fram Strait deep-water passage is given by Jakobsson et al. (2007) and Ehlers and Jokat (2008) with 17-18 Ma, but the beginning of the turbidity formation in our investigation area can be not dated, because of the limited seismostratigraphy control by deep drill holes in the GReenland Basin (Berger and Jokat, 2008). Accordingly, it is very speculative that all these events are results triggered by the opening of the deep-water gateway between the Arctic Ocean and the northern North Atlantic in interaction with the glacial onset and sea level changes in the Northern Hemisphere. In general, a shallow-water exchange through this passage might have started between 34 Ma and 25 Ma (Kristoffersen, 1990, Lawver et al., 1990, Jakobsson et al., 2007, Ehlers et al., submitted 2008), suggesting that the formation of the EGC started between 15-25 Ma. According to the palaeoceanographic model from Ehlers et al. (submitted 2008) the EGC should be active in water depths up to around 600 m. However, seismic data along the East Greenland margin from 71°N - 81°N (Berger and

Jokat, 2008; Berger and Jokat, submitted 2008) show no evidences in the upper 600 m (Ehlers et al., submitted 2008) for a massive sediment transport from north to south related to the EGC (Fig. ??). The reason can be tectonic structures like the Hovgård Ridge and Greenland Fracture zone. A branch turns to the east if the current come across these structures (Rudels et al., 2002) and the flow velocity of the north-south branch decreases and consequently also the erosional force (Fig. ??). For the sediment drift body observed on profile 7 (Fig. 7.4), we find an explanation in the modelled palaeoceanography for the North Atlantic (Ehlers et al., submitted 2008). Their oldest modelled time slice (20 Ma) shows a warm water transport (5.5°C) to the south. These water masses probably transported sediment towards the south and built out this local sediment mound in the slope area. This interpretation is supported by the base of the structure within unit GB-1 (older than middle Miocene). Additionally the sediment body is located a bit south of the greatest glacial trough on the Northeast Greenland shelf. Therefore, it is also possible that the material is coming from the East Greenland shelf and was transported by the EGC further south.

7.5.2 Deep-sea sedimentation in the northern North Atlantic

The influence of bottom currents on the sediment deposition and erosion at the seafloor is particularly strong in areas where high complex basement morphology led to a deflection or focusing of these currents and with it an intensification of the flow (Stoker, 1998). From the central Fram Strait up to 78°30′N a patch of fine-layered sedimentary strata is observed (Figs. 7.5: profiles 10-12, 7.6: profile 13) in water depths of more than 2500 m, and has been described as contourites (Berger and Jokat, submitted 2008); current-controlled sediments. In general, drift sediments were deposited away from the high-velocity core of currents (Howe et al., 1998). That would explain the sediment depocenters west and east of the seafloor depression on profile 13 (Fig. 7.6). The accumulation of sediments in terms of the flow direction is highly influenced by the Coriolis Force, which is triggered by the Earths rotation. On the Northern Hemisphere, currents will be deflected to the right side of their flow direction, which results in sediment accumulation to the right of the flow direction as well (Faugéres et al., 1999; Rebesco et al., 2002). Therefore, the sediment load at CDP 4000 (Fig. 7.6) is caused by deep-water export from the Arctic Ocean through the Fram Strait flowing in the southern direction. According to this, our interpretation support the palaeobathymetric modelling results from Ehlers et al. (submitted 2008) for the deep-water outflow through the Fram Strait for present day. They explain their results with a decreasing of salinity and temperature below 1500 m. At the western flank of a prominent basement high at CDP 5300 (Fig. 7.6) we also find a fine-bedded sediment succession. Due to the current transported material to the right side of the flow direction in the Northern Hemisphere, we believe in a northwards directed current in the eastern part of the channel (Fig. 7.6). Rudels et al. (2002) and Ehlers et al. (submitted 2008) postulate a present day northwards flowing West Spitsbergen Current entering the Arctic Ocean through the eastern Fram Strait (Fig. 7.7). Thus, it seems most likely to be that we have on the western flank of the basement high, West Spitsbergen Current transported sediment deposits.

In the modelling from Ehlers et al. (submitted 2008) we see no evidences for a deep-water outflow from the Arctic Ocean before 15 Ma. Our seismostratigraphic model on profile 13 (Fig. 7.6) shows the base of the levee structure west of the channel (CDP 4000) in the middle of unit II (3 - 7 Ma; Berger and Jokat, submitted 2008), which fits to an age of around 5 Ma. However, more current-related material (contourites) can be recognized between CDP 4000 and CDP 5300 below the deep-sea channel. Fine-laminated contourites are visible in depths up to 5000 ms (TWT) in the eastern part and up to 4700 ms (TWT) in the western part

7.5 Discussion

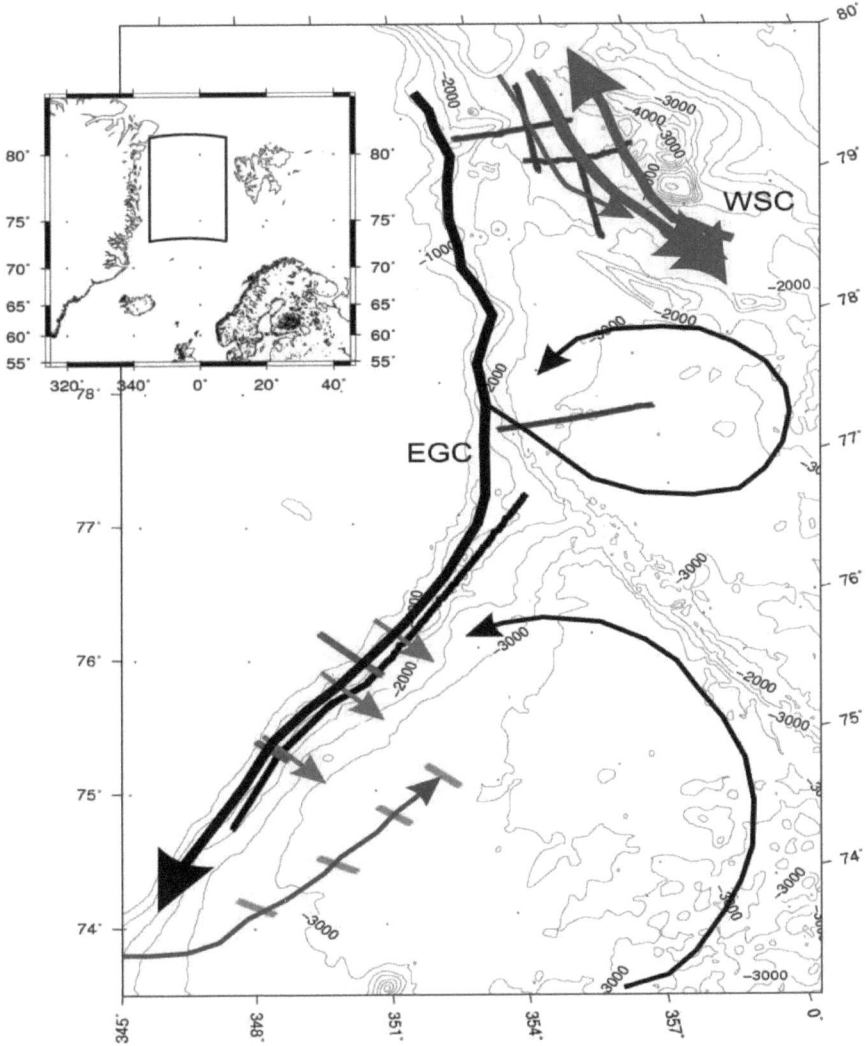

Figure 7.7: Interaction of different water masses in the North Atlantic for the study area with bathymetric contours in the background. Black arrows demonstrate current flow lines adopted from Rudels et al. (2002) and Ehlers et al. (submitted 2008). Red arrows show interpreted bottom circulations interpreted in this study. Blue arrows image alongslope directed turbidity currents identified in the slope region adjacent to the Greenland Basin. Comparatively thicker lines represent assumed stronger erosional character of the current. WSC = West Spitsbergen Current, EGC = East Greenland Current.

below the channel system. This support the interpretation that the deep-water outlflow of the Arctic Ocean (Fig. 7.6: CDP 4000 - 4600) is active since middle Miocene times (Berger and Jokat, submitted 2008) and the northwards directed West Spitsbergen Current (Fig. 7.6; CDP 4600 - 5300) probably since the opening of the deep-water connection between 17-18 Ma (Jakobsson et al., 2007, Ehlers and Jokat, 2008). The transport of sediments in Middle Miocene times through the Fram Strait was also assumed by Knies and Gaina (2008). In their interpretation, they postulate a sediment transport from the Barents Sea shelf and accumulation in the Molloy Basin, which seems to be not impossible judging from our latest results. The observed levee to the west of the channel (CDP 4000) in figure 7.6 is being interpreted as a result of southward flowing currents through the Fram Strait. The boundary between unit II and unit I, interpreted by Berger and Jokat (submitted 2008) presents an increase of deep-water outflow from the Arctic Ocean. This increase could again be triggered by an intensification of glacial and interglacial cycles in the Northern Hemisphere (Zachos et al., 2001).
The north-south profile 10 (Fig. 7.5), also located in the Molloy Basin shows a small depression at the northern end. However, the depression is just visible on the seafloor and does not continue into deeper layers. Therefore, it seems to be a young event, however, a higher sediment load north or west of this channel could not be observed. This profile crosses profile 13 (Fig. 7.6: CDP 3500), west of the sediment depocenter which is caused by Arctic Ocean deep-water outflow. It seems to be more likely a west-east flowing current has produced the depression of the seafloor on profile 10 (Figs. 7.5, 7.7).
The profile in the central part of the Boreas Basin (Fig. 7.5: line 9) shows a completely different sediment accumulation in the deep sea to the northern Molloy Basin. The nearly transparent sediment facies do not point to current-related sedimentation. However, Ehlers et al. (submitted 2008) published in their article, that the outflow of the Arctic Ocean crosses the Boreas Basin and merges with the re-circulating Atlantic Water in present times (Fig. 7.7). In general, fracture zones and ridges can influence also the flow direction of deep sea currents (Rudels et al., 2002). The Hovgård Ridge and the Greenland Fracture zone in the North Atlantic may represent such barriers. Currents coming from the north were probably partly deflected to the eastern direction (Fig. 7.7). According to that, the channel on profile 10 may have been built by re-circulating Atlantic Water or deflected Arctic Water north of the Hovgård Ridge.

7.5.3 Deep-sea sedimentation in the central Greenland Basin

A channel system with increasing sediment accumulation on the eastern flank has been observed in the deep Greenland Basin. Seismic profile 1 (Fig. 7.2) demonstrates the best preserved elongated drift body with a sediment thickness of around 170 ms (TWT), which decreases in the northern direction. As already mentioned, the Coriolis Force in the Northern Hemisphere has an influence on the accumulation of sediments, stimulating deposition on the right hand side of the current flow direction (Faugéres et al., 1999). If we believe in current controlled sedimentation we would expect now a flow direction of the current towards the north, forming the observed channel-levee structure in the deep Greenland Basin (Fig. 7.7). However, oceanographic results from Rudels et al. (2002) and the palaeoceanographic modelling from Ehlers et al. (submitted 2008) show a major flow direction (EGC) along the East Greenland margin from north to south and not in the opposite direction. Otherwise we can not recognize a prolongation of this channel north of 75.2°N and south of 74°N and therefore it seems to be very local, and untypical for a current induced channel system. In general, the area between 72°N and 75°N is extensively studied in terms of submarine channels. Investigations (Dowdeswell et al., 2002; RV Polarstern in 2000, 2001 and 2002; Lemke, 2003; Cofaigh et al., 2004) show four main channel systems in this area in different water depth and with different dimensions. Sediment cores demonstrate bioturbated sandy to silty clays at the surface underlain by laminated silts and clays which may indicate that the channels are in an inactive phase today. Retrieved sediment cores from the levees show distinct changes from laminated to crossbedded units consisting of clays to silts and bioturbated silty clays, which reflects episodic deposition of distal turbidities or bottom current activity (Lemke, 2003). The origin of the channel system is still under discussion. Mienert et al. (1993) and Dowdeswell et al. (2002) proposed that the channels in the area between 72°N and 75°N relate to downslope flow of dense-water and turbidity currents originating from sea-ice formation on the shelf and upper slope. Wilken and Mienert (2006) assume major glaciations with varying ice-stream behaviours for the formation of the channels. The channel system in the deep Greenland Basin is located in the prolongation of the Ardencaple Fjord, which forms a large glacial trough system on the continental East Greenland shelf (Berger and Jokat, submitted 2008). The small depression of the channel and the almost undisturbed sediments in the upper part of unit GB-2 led to the assumption that the turbidity channel has developed in the last few million years, also supported by Wilken and Mienert (2006). Berger and Jokat interpret the base of unit GB-2 as onset of glaciation on the Northeast Greenland shelf. We see the transition from well-laminated sediments to the chaotic sedimentary strata within the middle of the unit GB-2 section on these profiles, which might be an indication for intensification of glacial and interglacial cycles and the development of turbidity activity. The flow direction of the turbidity current turns towards the north down to the abyssal plain (Fig. 7.7). The existing sediment cores are not deep enough to perform an age for the beginning of the formation of the turbidity channel. Compared to the channel system in the deep Molloy Basin (Fig. 7.6) and a similar change from thin-bedded sediments to a chaotic sediment strata we assume the same formation age of $\sim$5 Ma.

7.6 Conclusion

In the area between 75°N and 81°N, we find different current induced sediment deposits in the deep sea and in the slope region on the East Greenland margin. Especially in the deep Molloy Basin, we found a fine-layered sediment character which we interpret as contourite deposits. These sediments can be also observed west and east of a channel system, which indicates the deep-water outflow and inflow from and to the Arctic Ocean and leads to sediment levees west and east of the channel. The seismostratigraphy induced an age of 5 Myrs for the base of the western levee structure, probably influenced by an intensification of glacial and interglacial cycles to this time. Below the channel system up to the basement, a well-laminated sediment succession is visible which leads to the assumption, that current induced sedimentation is active since the formation of this passage. Another channel system is present in the deep Greenland Basin and can be interpreted as turidity channel formed during major glacial times and probably existing since 5 Ma. The slope region adjacent to the Greenland Basin is influenced by downslope turbidity currents which lead to ripple formations identified on slope parallel profiles and interpreted as turbidity channels. Also a slump structure and sediment drift body were interpreted in this dataset and the seismostratigraphy shows that all the events on the slope adjacent to the Greenland Basin have their origin before 15 Ma, because they could be traced down into unit GB-1. Possible potential trigger events are variable sediment input as a response to advancing/retreating ice sheets during glacial cycles. Material was transported across the shelf and transported towards the deep-sea.

7.7 Acknowledgements

The authors thank the captain and crew of the German Research Vessel "Polarstern". They thank also Berit Kuvaas at StatoilHydro ASA, for useful comments and suggestions. This project was partly funded by the Deutsche Forschungsgesellschaft and StatoilHydro ASA.

7.8 References

Berger, D. and Jokat, W., (2008). *A seismic study along the East Greenland margin between 72 - 77°N*. Geophysical Journal International, Vol. 174 (2), page: 733-748.

Berger, D. and Jokat, W., (submitted 2008). *Sediment deposition in the northern basins of the North Atlantic and characteristic variations in shelf sedimentation along the East Greenland margin*. Marine and Petroleum Geology.

Brancolini, G., Cooper, A. K., & Coren, F., (1995). *Seismic Facies and Glacial History in the Western Ross Sea (Antarctica)*. Antarctic Research Series. American Geophysical Union, Washington, DC, Vol. 68, pp. 209-234.

Bourke, R. H., Paquette, R.G., Blythe, R.F., (1992). *The Jan Mayen Current of the Greenland Sea*. Journal of Geophysical Research, Vol. 97, No. C5, pp. 7241-7250.

DeSantis, L., Brancolini, G. and Donda, F., (2003). *Seismostratigraphic analysis of the Wilkes Land continental margin (East Antarctica): Influence of glacially driven processes on the Cenozoic deposition*. Deep-Sea Res. II 50, 1563-1594.

Dowdeswell, J.A. and O Cofaigh, C. (2002). *Glcier-influenced sedimentation on high-latitude continental margins: introduction and overwiew*. Geological Society, London, Special Publication, 203, 1-9.

Dowdeswell, J.A., O Cofaigh, C., and Pudsey, C.J., (2004). *Thickness and extent of the subglacial till layer beneath an Antarctic paleo-ice stream*. Geology, v. 32, p. 13-16.

Ehlers, B.-M. and Jokat, W., (2008). *Analysis of subsidence in crustal roughness for ultra-slow spreading ridges in the northern North Atlantic and the Arctic Ocean*. Geophysical Journal International, Vol. 177, Issue 2, pp. 451-462.

Ehlers, B.-M., Butzin, M., Grosfeld, K. and Jokat, W., (submitted 2008). *A palaeoceanographic modelling study of the Cenozoic northern North Atlantic and the Arctic Ocean*. Global and Planetary Change.

Eldrett, J.S., Harding, I.C., Wilson, P.A., Butler, E., Roberts, A.P., (2007). *Continental ice in Greenland during the Eocene and Oligocene*. Nature, Vol. 446, 176-179.

Escutia, C., Warnke, D.A., Acton, G.D., Barcena, A., Burckle, L., Canals, M., Frazee, C.S., (2003). *Sediment distribution and sedimentary processes across the Antarctic Wilkes Land margin during the Quaternary.* Deep-Sea Research. Part 2. Topical Studies in Oceanography 50, 1481- 1508.

Faugéres, J.-C., Stow, D.A.V., Imbert, P., Viana, A., (1999). *Seismic feature diagnostic of contourite drifts.* Marine geology 162. 1-38.

Forsberg, C.F., Solheim, A., Elverhi, A., Jansen, E., Channell, J.E.T., Andersen, E.S., (1999). *The depositional environment of the western Svalbard margin during the upper Pliocene and the Pleistocene; sedimentary facies changes at Site 986.* In: Raymo, M.E., Jansen, E., Blum, P., Herbert, T.D. (Eds.), Proceedings of the Ocean Drilling Program. Scientific Results, vol. 162. Ocean Drilling Program, College Station, TX, pp. 233-246.

Howe, J.A., Livermore, R.A., Maldonado, A., (1998). *Mudwave activity and current-controlled sedimentation in Powell Basin, northern Weddell Sea, Antarctica.* Marine Geology 149. 229-241.

Jakobsson, M. et al., (2007). *The early Miocene onset of a ventilated circulation regime in the Arctic Ocean.* Nature, Vol. 447, 986-990.

Knies, J. and Gaina, C., (2008). *Middle Miocene ice sheet expansion in the Arctic: Views from the Barents Sea.* Geochemistry Geophysics Geosystems, 9, Q02015, doi:10.1029/2007GC001824.

Kuvaas, B. and Leitchenkov, G., (1992). *Glaciomarine turbidite and current controlled deposits in Prydz Bay, Antarctica.* Mar. Geol. 108, 365-381.

Lawver, L.A., and Gahagan, L.M., (2003). *Evolution of Cenozoic seaways in the circum-Antarctic region.* Palaeogeography, Palaeoclimatology, Palaeoecology, 198, 11-37.

Lemke, P. (Ed.), 2003. *The Expedition ARKTIS XIII/1 a, b of the Research Vessel Polarstern in 2002.* Rep. Pol. Mar. Res. 446, 120 pp.

Mienert, J., Kenyon, N.H., Thiede, J., Hollender, F.-J., 1993. *Polar continental margins: studies off East Greenland.* EOS, Trans. Am. Geophys. Union 74, 225, 234, 236.

Miller, H., Henriet, J. P., Kaul, N., and Moons, A., (1990). *A fine scale seismic stratigraphy of the eastern margin of the Weddell Sea.* In: Bleil, U., Thiede, J. (Eds.) Geological History of the Polar Oceans: Arctic versus Antarctic. Kluwer Academ. Publ., Boston, 131-161.

Myhre, A. M., Thiede, J., Firth, J. V. and Shipboard Scientific Party, (1995). *Initial Reports: sites 907-913, North Atlantic-Arctic Gateways I.* Proceedings, Initial Reports, Ocean Drilling Program, ODP 151: 926 pp. (Ocean Drilling Program, College Station, TX).

Rebesco, M., Pudsey, C. J., Canals, M., Camerlenghi, A., Barker, P. F., Estrada, F., Giorgetti, A., (2002). *Sediment drifts and deep-sea channel systems, Antarctic Peninsula*

Pacific Margin. Geological Society Memoirs, 22, 353 - 372.

Rise, L., Ottesen, D., Berg, K., Lundin, E., (2005). *Large-scale development of the mid-Norwegian margin during the last 3 million years.* Marine and Petroleum Geology 22, 33-44.

Rudels, E., Fahrbach, E., Meincke, J., Budus, G., Eriksson, P., (2002). *The East Greenland Current and its contribution to the Denmark Strait overflow.* Journal of Marine Science, 59: 1133-1154.

Stein, R., 2003. Developments in marine Geology: Arctic Ocean sediments: Processes, Proxies, and Palaeoenvironment.

Stemmerik, L. et al., (1993). *Depositional history and petroleum geology of Carboniferous to Cretaceous sediments in the northern part of East Greenland.* In: Vorren, T.O. et al., (Eds.), Arctic Geology and Petroleum Potential. Norwegian Petroleum Society (NPF), Special Publication 2. Elsevier, Amsterdam, pp. 67-87.

Stoker, M.S., Akhurst, M.C., Howe, J.A., Stow, D.A.V., (1998). *Sediment drifts and contourites on the continental margin off northwest Britain.* Sedimentary Geology 115, 33-51.

Stoker, M.S., Akhurst, M.C., Howe, J.A., Stow, D.A.V., 1998. *Sediment drifts and contourites on the continental margin off northwest Britain.* Sedimentary Geology 115, 33-51.

Wilken, M., Mienert, J., 2006. *Submarine glacigenic debris flows, deep-sea channels and past ice-stream behaviour of the East Greenland continental margin.* Quaternary Science Reviews 25, 784-810

Wong, H.K., Ldmann, T., Baranov, B.V., Karp, B.Ya., Konerding, P., Ion, G., (2003). *Bottom current-controlled sedimentation and mass wasting in the northwesten Sea of Okhotsk.* Marine Geology 201, 287-305.

Zachos, J., Pagani, M., Sloan, L., Thomas, E., Billups, K., (2001). *Trends, Rhythms, and Aberrations in Global Climate 65 Ma to Present.* Science 292, 686-693.

8 Conclusion

This study is engaged in the understanding of the sedimentation history in terms of glaciation, sediment accumulation during and before glacial times and in ocean circulations which trigger mass transport within deep sea and slope regions. The analysis of the sedimentation history along the East Greenland margin and the northern North Atlantic is possible due to new seismic data gathered by the Alfred Wegener Institute (AWI) during the last years. The data provide for the first time a detailed view into the sediment structure and provide a basis for age estimations of the sediments. The following section concludes the results of this study, presented in the chapters 5, 6 and 7, according to the four main questions.

1. How did the passive continental margin of East Greenland develop after the opening of the North Atlantic?

   The opening of the North Atlantic took place 54 Ma, with the initial opening in the Greenland Basin, which represents the oldest basin in the North Atlantic. During the initial phases of the opening of the Greenland Basin a prominent, diachronuously continuously structure with a N-S extent of approximately 360 km has been developed. This volcanic structure influences the sedimentation into the deep Greenland Basin. The different opening times of the three northernmost basins (Molloy Basin, Boreas Basin, Greenland Basin) induce variations in the accumulation of sediments compared from north to south. The formation of continental fragments, like the Hovgrd Ridge and the Greenland Fracture zone contributes these sedimentary variations. Differences in sedimentation along the East Greenland margin are also visible after the onset of glaciation. The interaction between the topography of the Greenland mainland and various ice sheet flow directions led to different developments of the Greenland shelf and the adjacent slope region. A narrow shelf with a steep slope on the Southeast Greenland margin changes into a wide shelf with a gently dip of the slope towards north. A different climatic regime with stronger and thicker ice package in the north is remains very speculative.

2. How often did the ice sheets advance to the shelf break and is it possible to give an estimate about the onset of the glaciation through an age correlation?

   The ODP drill sites 909 and 913 provide age information in the deep sea area of the Molloy and Greenland Basin. The seismostratigraphic correlations into these areas give different ages for the onset of progradation, which will be interpreted as sea level changes and/or glacial erosion by an early ice sheet or glaciers. The begin of the progradation within the shelf area adjacent to the Greenland Basin can be clearly differentiated from aggradational units below, and is dated to be of Middle Miocene age. A subdivided classification of aggradational strata and progradational strata could be not carried out on the Molloy Basin shelf. The data show a limited view on the deeper sediment structure within the shelf region. A prominent reflector and simultaneously the last clearly identified horizon belong to the progradational unit and has an age of middle Late Miocene. It is still unclear if this reflector marks the onset of glaciation in that region.

Conclusion

In general, the prograding strata north of 67°N could be dicided into two units (SU3 = Middle Miocene to middle Late Miocene, SU2 = middle Late Miocene to Pliocene) on the basis of seismostratigraphy and a comparison of the sediment character on the Southeast Greenland margin. The limited age information of the boreholes and the limited seismic solution does not allow a more detailed classification of the prograding sequences and therefore it is difficult to give a statement about the advances of the ice sheets. On the other hand, a differentiation between sea level related progradation and ice sheet advancing related progradation is not possible.

3. Are there lateral variations in the sedimentation along the East Greenland margin, and if so, what are the influencing factors?

- shelf region:

  The shelf region shows strong differences in the width of the shelf. The greatest extension since the onset of glaciation is visible between 70°N and 77°N in the prolongation of glacial troughs with up to 60 km. The most material will be transported along these troughs, which seems to have the main influence on the accumulated sediments on the shelf. The glacial troughs developed in interaction with the direction of ice sheets movements. Additionally, the rough topography in the south of Greenland leads to more less sediment transport to the west, observable also on the narrower shelf and steep slope in comparison with the area north of 70°N.

- slope region:

  Within the slope, adjacent to the Greenland Basin a prominent 360 km long volcanic structure was observed, strongly influencing the sediment transport from the shelf into the deep sea area. North of the Greenland Basin the basement structure is missing in the slope area, so that a continuous sediment accumulation could take place. Furthermore, also turbidite activity was observed just on the slope adjacent to the Greenland Basin. Maybe, caused by the increased sediment transport along the glacial troughs. These interpreted tubidite structures can be traced down up to the Middle Miocene unconformity. On the Southeast Greenland margin at around 70°N, a lot of sediment material is accumulated in the prolongation of the Scoresby Sund. Probably, the East Greenland Current coming from the north transported these sediments further southwards and build out some current-related sediment packages.

- deep sea region:

  The greatest variations can be found in the deep sea areas. Sediment thickness maps image the variation in the sedimentation along the East Greenland margin and show the lowest total sediment thickness in the Greenland Basin with around 1000 m. That is the effect of the volcanic structure located within the slope and acting as barrier for sediment transport from the shelf. Differences were also observed in terms of the sediment character. In the Molloy Basin and Greenland Basin well-laminated levee structures were found, caused by current activity. However, the Boreas Basin represents a chaotic sediment facies and the sedimentation seems to be not influenced by current-controlled deposition.

4. Can seismically observable sediment structures give information about current activity and climate changes in the region investigated?

Current-controlled sedimentation could be observed south of the Fram Strait, within the slope area, and as well as in the deep Greenland Basin. Due to the geometry of the sediment accumulation an outflow of the Arctic Ocean through the western Fram Strait and an inflow into the Arctic Ocean trough the eastern Fram Strait could be identified. The sediment deposits west an east of the channel will be interpreted as levee structures. The seismostratigraphy induced an age of 5 Myrs for the base of the western levee structure, probably influenced by an intensification of glacial and interglacial cycles to this time. Deeper well-laminated sediments leads to the assumption that current activity is active since the formation of the Fram Strait deep-water passage which fits with first evidences for a glacial stage. Additionally, we find turbidite activity in the slope region of the Greenland Basin down to sediment older than 15 Ma. Probably, trigger events for this sediment transport are advances and retreats which trigger an increased sediment transport towards the shelf edge into the slope.

Conclusion

9 Acknowledgements

I thank Prof. Dr. Heinrich Miller my supervisor for the allocation of this work and the possibility to analyze the data set with an excellent equipment of the Alfred Wegener Institute. Thanks also to Prof. Dr. Cornelia Spiegel for the interest in this work and the willingness to co-reviewing my work.

A great thank goes to Dr. Wilfried Jokat my motivator and principal supporter. Thanks for the scientific and many many not-so-scientific discussions, funny trips to different meetings and controversial discussions about men and women, especially about the three German Ks.

I am grateful to Dr. Erik Lundin and Bruce Alastair Tocher for giving me an exciting stay in Stavanger to get a first introduction in the work of an international exploration company. Especially, Dr. Erik Lundin I thank also for his helpful comments and English improvements of my manuscript. Tack sa mycket, Speedy Gonzales!

Many thanks goes to my colleagues in the Marine Geophysical working group, especially to Dr. Wolfram Geissler, Dr. eStella Weigelt, Dr. Mechita Schmidt-Aursch, Dr. Catalina Gebhardt and Dr. Max Voss for helpful comments, discussions and improving the manuscript.

Dr. Carsten Scheuer and Birte-Marie Ehlers I thank for the very funny and friendly atmosphere in our office. I will never forget you.

I would like to thank also the Deutsche Forschungsgesellschaft and StatoilHydro ASA for the financial support.

A great thank goes to my parents for the support and understanding, as well as the full chocolate supply every time at home.

Finally, the greatest thank goes to Christian, for his patience and motivation all the time, to always being there for me and and and...

Die VDM Verlagsservicegesellschaft sucht für wissenschaftliche Verlage abgeschlossene und herausragende

Dissertationen, Habilitationen, Diplomarbeiten, Master Theses, Magisterarbeiten usw.

für die kostenlose Publikation als Fachbuch.

Sie verfügen über eine Arbeit, die hohen inhaltlichen und formalen Ansprüchen genügt, und haben Interesse an einer honorarvergüteten Publikation?

Dann senden Sie bitte erste Informationen über sich und Ihre Arbeit per Email an *info@vdm-vsg.de*.

Sie erhalten kurzfristig unser Feedback!

VDM Verlagsservicegesellschaft mbH
Dudweiler Landstr. 99
D - 66123 Saarbrücken
www.vdm-vsg.de

Telefon +49 681 3720 174
Fax   +49 681 3720 1749

Die VDM Verlagsservicegesellschaft mbH vertritt

Printed by Books on Demand GmbH, Norderstedt / Germany